SpringerBriefs in Physics

SpringerBriefs in Physics are a series of slim high-quality publications encompassing the entire spectrum of physics. Manuscripts for SpringerBriefs in Physics will be evaluated by Springer and by members of the Editorial Board. Proposals and other communication should be sent to your Publishing Editors at Springer.

Featuring compact volumes of 50 to 125 pages (approximately 20,000–45,000 words), Briefs are shorter than a conventional book but longer than a journal article. Thus, Briefs serve as timely, concise tools for students, researchers, and professionals.

Typical texts for publication might include:

- A snapshot review of the current state of a hot or emerging field
- A concise introduction to core concepts that students must understand in order to make independent contributions
- An extended research report giving more details and discussion than is possible in a conventional journal article
- A manual describing underlying principles and best practices for an experimental technique
- An essay exploring new ideas within physics, related philosophical issues, or broader topics such as science and society

Briefs allow authors to present their ideas and readers to absorb them with minimal time investment. Briefs will be published as part of Springer's eBook collection, with millions of users worldwide. In addition, they will be available, just like other books, for individual print and electronic purchase. Briefs are characterized by fast, global electronic dissemination, straightforward publishing agreements, easy-to-use manuscript preparation and formatting guidelines, and expedited production schedules. We aim for publication 8–12 weeks after acceptance.

Alejandro Algora · Berta Rubio · Jose Luis Tain ·
William Gelletly

Total Absorption Technique
for Nuclear Structure
and Applications

 Springer

Alejandro Algora
Instituto de Fisica Corpuscular (IFIC)
CSIC-University of Valencia
Paterna (Valencia), Spain

Jose Luis Tain
Instituto de Fisica Corpuscular (IFIC)
CSIC-University of Valencia
Paterna (Valencia), Spain

Berta Rubio
Instituto de Fisica Corpuscular (IFIC)
CSIC-University of Valencia
Paterna (Valencia), Spain

William Gelletly
Department of Physics
University of Surrey
Guildford, UK

ISSN 2191-5423   ISSN 2191-5431 (electronic)
SpringerBriefs in Physics
ISBN 978-3-031-58863-1   ISBN 978-3-031-58864-8 (eBook)
https://doi.org/10.1007/978-3-031-58864-8

This Springer imprint is published by the registered company Springer Nature Switzerland AG
The registered company address is: Gewerbestrasse 11, 6330 Cham, Switzerland

If disposing of this product, please recycle the paper.

Preface

This book presents the first detailed description of the use of the total absorption technique in beta-decay studies.

When a nucleus undergoes beta decay, the beta-decay probability to a particular level in the daughter nucleus provides basic nuclear structure information that can be used for testing nuclear models and probing our knowledge of the nuclear properties of both parent and daughter states. This information is not only useful for basic nuclear structure research; it is also relevant for nuclear astrophysics and for many practical applications like the prediction of the reactor decay heat and the prediction of the primary antineutrino spectrum from nuclear reactors. In conventional high-resolution beta-decay studies the beta-decay probability to a level is determined indirectly, by first constructing a level scheme populated in the daughter nucleus and then extracting the beta-decay probabilities from the gamma intensity balance of the different levels. How well those probabilities are determined depends on the efficiency of the used high-resolution setup, which is composed in general of several Ge detectors. If the beta decay has a large Q value, or it is very complex and populates regions in the daughter nucleus with high-level density, it is very probable that some de-exciting gamma rays remain undetected. This is a systematic error, which causes assigning larger beta-decay probabilities than the real ones to the lower lying levels. This systematic error is commonly called the Pandemonium effect.

Nowadays, there is only one accepted technique that can provide nuclear decay data free from the Pandemonium effect. This technique, the total absorption technique, is based on the detection of the gamma cascades that follow the beta decay using highly efficient devices. Total absorption devices are constructed using scintillator material of high density and volume covering the sources in approximately 4π geometries. This book will cover all the relevant topics related to the use of this technique, from the preparation of an experiment to the algorithms of analysis. It covers also the most relevant applications of the technique, from nuclear structure (comparison with nuclear models, determination of nuclear shapes) to practical applications (reactor decay heat, prediction of the antineutrino spectrum from reactors), and astrophysics.

This book will be an excellent tool for scientists and advanced students interested in beta-decay studies as well as in practical applications. It provides a detailed introduction to the beta-decay process and all the necessary know-how for the use of the total absorption technique. The presented algorithms of analysis and techniques are examples of solutions of the inverse problem that can also be of interest for applications in other disciplines.

Paterna (Valencia), Spain   Alejandro Algora

Acknowledgements

The results presented in this book were only possible thanks to the dedicated contribution of many Ph.D. students, colleagues, and technicians who have collaborated in this work.

Contents

Part III   Applications and Future

Part I
Introduction to the Total Absorption
Gamma Spectroscopy Technique

Chapter 1
Introduction to Beta Decay

Abstract In this chapter the basic concepts of the beta decay process will be presented. First a brief historical background is given. Later the basic relations of beta decay and the Fermi theory of beta decay are introduced. The beta decay of ^{24}Na is then presented as an example. Finally the *Pandemonium* effect is introduced.

1.1 Early History

Radioactivity was discovered by chance in Uranium salts by Becquerel [1] in 1896. Later he [2], and independently Kaufmann [3], showed that the radiation, whose effects he had seen, is due to electrons. Meanwhile his work was followed up by Marie Curie and Pierre Curie, who showed that other radioactive species existed, namely Thorium and Radium. Shortly afterwards, Rutherford [4] was able to identify two types of radioactive emission that were clearly different judging by their degree of penetrability in objects and their ability to cause ionisation. Rutherford named them alpha and beta, using the first letters of the Greek alphabet. Another type of radioactivity was soon added, after the identification by Villard [5] of another, even more penetrating radiation, later also confirmed by Rutherford and named gamma. All these discoveries paved the way to the nuclear model of the atom proposed later by Rutherford [6], based on the work of Geiger and Marsden [7]. In essence the model consists of a positively charged heavy nucleus surrounded by orbiting electrons. The radiations observed were emitted by the nucleus and their nature identified as helium nuclei (α particles), electrons (β radiation) and high energy photons (γ radiation). In 1901 Soddy and Rutherford [8] showed that alpha and beta radioactivity are related to the transmutation of chemical elements.

The basic structure of the nucleus in terms of its constituents was finally settled in 1932 with the discovery of the neutron by Chadwick [9]. Since then we believe that nuclei are composed of Z protons and N neutrons. The sum of the protons

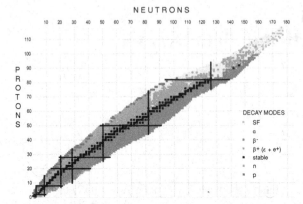

Fig. 1.1 Our present knowledge of the nuclide chart, or Segrè chart as it is known. In essence this table is the nuclear physics equivalent of the Mendeleev table in chemistry. Each square on the table represents an isotope. Stable or very long-lived nuclei are represented in black. Blue and red colours show nuclei that can undergo beta decay. The lines show the position of the magic numbers. The "magic numbers" correspond to proton and neutron numbers that characterise systems of increased energetic stability. The x axis represents the number of neutrons and the y axis the number of protons. Courtesy of Filip Kondev, using data from NUBASE2020 evaluation [10]

and neutrons gives us the mass number ($A = N + Z$) of the nucleus. Isotopes are characterised by having the same Z, so they behave chemically in the same way but have different numbers of neutrons (N).[1]

Among the different forms of radioactivity, beta decay is special. Nowadays we know that it is the most general form of element transmutation in the nuclide chart. In Fig. 1.1, we can see the presently known nuclear species, of which there are about 3600. Among them, only a limited number is stable, or have very long half-lives compared with human existence. The remainder exhibit different forms of radioactive decay, with β decay being the most common.

The spectra of α- and γ-rays consist of discrete lines with well defined energies. In contrast the beta spectra are continuous, something that was pointed out by Chadwick [11] at an early stage. The interpretation of the continuous character of the spectrum caused serious difficulties in the early years of the 20th century, because it was at odds with energy conservation. The discrete nature of α and γ radiation was interpreted as the result of radiation being emitted between discrete nuclear states, similar to the electromagnetic radiation emitted in transitions between excited states of the atom. So, how can a continuous spectrum be interpreted in the new subatomic world dominated by quantum transitions? The puzzling phenomenon of β decay even led to suggestions by Niels Bohr that the energy conservation was only valid in a statistical sense. The solution came from Pauli [12], who suggested that in the process of beta decay a light neutral particle is also emitted, which shares the energy released and

---

[1] Please note that the chemistry of the elements (atoms) is determined by the number of electrons in the atomic shells. Electrical neutrality of the atom as a whole implies that the number of electrons in the atomic shells is the same as Z.

linear momentum with the emitted electron. He named the particle *neutron*, a name that was later changed to *neutrino* ('little neutral one' in Italian) as suggested by Enrico Fermi, in his theory of beta decay [13].

1.2 Basic Relations

In this section we will review the various beta decay processes. At the level of the nucleons, the three possible processes are:

$$
\begin{aligned}
n &\rightarrow p + e^- + \bar{v} \quad \textit{beta minus decay} \ (\beta^-) \\
p &\rightarrow n + e^+ + v \quad \textit{beta plus decay} \ (\beta^+) \\
p + e^- &\rightarrow n + v \quad \textit{electron capture} \ (\varepsilon)
\end{aligned}
\tag{1.1}
$$

In essence beta decay is the spontaneous transformation of an unstable nucleus into an isobar, a nucleus with the same mass number A, with the emission of one electron(positron) or the capture of one electron from the atomic shell. In all three processes a neutral particle is also emitted. In β^- decay it is an antineutrino, $\bar{v}$, and in the β^+ and electron capture (EC) processes it is a neutrino, v.

In β^- decay, a neutron is converted into a proton, so the process is energetically possible when

$$
M(A, Z) > M(A, Z + 1) + m_e
\tag{1.2}
$$

where $M(A, Z)$ represents the mass of the nucleus with mass number A and atomic number Z, and m_e is the electron mass. In terms of atomic masses (after adding $Z * m_e$ to both sides of Eq. 1.2), which are tabulated [14], we obtain:

$$
M_{at}(A, Z) > M_{at}(A, Z + 1)
\tag{1.3}
$$

as the necessary condition for the β^- decay to occur.

Similarly for the β^+ decay process we can write:

$$
M(A, Z + 1) > M(A, Z) + m_e
\tag{1.4}
$$

after adding $(Z + 1) * m_e$, for β^+ decay

$$
M_{at}(A, Z + 1) > M_{at}(A, Z) + 2m_e
\tag{1.5}
$$

and for the electron capture process (ε):

$$M(A, Z + 1) + m_e > M(A, Z) \tag{1.6}$$

after adding $Z * m_e$, we obtain

$$M_{at}(A, Z + 1) > M_{at}(A, Z) \tag{1.7}$$

It should be noted that when the vacancy in the atomic shell, created in the absorption of the electron in the atomic shell, is filled one observes X-rays. This led to the discovery of the process by Alvarez [15]. This form of decay becomes more relevant for heavy nuclei, with larger Z values, because of the increasing physical overlap of the electron atomic shell wave functions with the nucleus.

Looking at relations (1.3), (1.5) and (1.7) and considering that in general $M(A, Z)_{at} \neq M(A, Z \pm 1)_{at}$ it is evident, that some form of beta decay is possible until we reach the minimum mass corresponding to a particular mass number A. It should be noted that so far we have only taken into account the constraints on the energy, but additional aspects of the decay, such as the quantum properties (spins and parities) of the states involved can also play a role. This will be discussed later.

In terms of energy it is important to consider the energy released, the Q value, in beta decay. This can be calculated as it is done conventionally in nuclear reactions. Thus for a β^- decay,

$$_{Z}^{A}X_N \rightarrow _{Z+1}^{A} X_{N-1}^* + e^- + \bar{\nu} \tag{1.8}$$

$$Q_{\beta^-} = [M(_{Z}^{A}X_N) - M(_{Z+1}^{A}X_{N-1}) - m_e]c^2 \tag{1.9}$$

(neglecting the $\bar{\nu}$ mass $m_{\bar{\nu}}$) which in terms of atomic masses becomes:

$$Q_{\beta^-} = [M_{at}(_{Z}^{A}X_N) - M_{at}(_{Z+1}^{A}X_{N-1})]c^2 \tag{1.10}$$

This energy can be shared by the electron and the antineutrino taking part in the decay:

$$Q_{\beta^-} = T_e + T_{\bar{\nu}} \tag{1.11}$$

where T represents the kinetic energy. It should be noted that the recoilling residual nucleus also has a small kinetic energy.

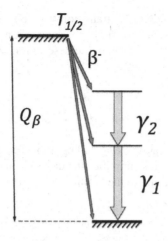

Fig. 1.2 Representation of a simple beta decay populating excited states in the daughter nucleus. The Q value is the mass difference between the ground state of the parent nucleus and the ground state of the daughter nucleus (see text for details). The states populated in the daughter nucleus can de-excite with the emission of gamma rays or by internal conversion when the energy of the transition is given to an electron of the atomic shell. The latter will lead to vacancies in the relevant atomic shell and this can be followed by X-ray emission or Auger electrons. If a level is long enough lived, it may also decay by β-decay

Similarly for the β^+ and electron capture processes the Q values are:

$$Q_{\beta^+} = [M_{at}(^A_{Z+1}X_N) - M_{at}(^A_Z X_{N+1}) - 2m_e]c^2 \tag{1.12}$$

$$Q_\varepsilon = [M_{at}(^A_{Z+1}X_N) - M_{at}(^A_Z X_{N+1})]c^2 - B_n \tag{1.13}$$

where B_n represents the binding energy of the captured electron.

Conventionally the Q value represents the maximum available energy for the decay process (see Fig. 1.2), so typically a beta transition from the ground state of the parent nucleus to the ground state of the daughter nucleus, but in the decay process it is common to populate excited states in the daughter nucleus. The energy shared by the particles is then less than the Q value.

1.3 Fermi Theory

After the introduction of the neutrino hypothesis by Pauli, Fermi was able to formulate a theory that explained the main features of the beta spectrum [13]: The theory used perturbation theory and assumed that the interaction was similar to the well-known electromagnetic interaction. In this chapter we will provide a simplified version of the theory for completeness.

We shall assume that we would like to calculate the transition rate from an initial state ψ_i to a final state ψ_f, where ψ represents the wave function of the given state.

For simplicity we will consider a β^- transition from a neutron single particle state to a proton single particle state, but similar considerations can be employed in β^+ decay. The final state ψ_f should contain the description of the fermions created in the process:

$$\psi_i = \psi_n(\vec{r}_n) \tag{1.14}$$
$$\psi_f = \psi_p(\vec{r}_p)\psi_e(\vec{r}_e)\psi_\nu(\vec{r}_\nu)$$

Here the recoil energy is assumed to be negligible and the spins of the particles are not taken into account. This simplifies the description considerably. The rate of the transition is given according to the Fermi's golden rule:

$$\lambda = \frac{2\pi}{\hbar}|H_{if}|^2 n_f(E_0) \tag{1.15}$$

where H_{if} is the matrix element of the operator H connecting the initial and final states and $n_f(E_0)$ represents the density of final states for a transition energy E_0.

$$H_{if} = \int \psi_f^* H \psi_i \, d^3\vec{r}_n \, d^3\vec{r}_p \, d^3\vec{r}_e \, d^3\vec{r}_\nu \tag{1.16}$$

The Hamiltonian H can take different forms, but in the simplest form it can be considered a contact interaction, which reflects the fact that the weak interaction is mediated by a particle of large mass, the W boson as we know it today. In this schematic model the Hamiltonian is just a constant proportional to the strength of the interaction (more rigorously $H = G\delta(\vec{r}_p - \vec{r}_n)\delta(\vec{r}_n - \vec{r}_e)\delta(\vec{r}_e - \vec{r}_\nu)$).

For the description of the electron (positron) and the antineutrino (neutrino) we might use plane waves of the form:

$$\psi_e = \frac{1}{V^{1/2}}e^{i\vec{k}_e\vec{r}_e} \tag{1.17}$$
$$\psi_\nu = \frac{1}{V^{1/2}}e^{i\vec{k}_\nu\vec{r}_\nu}$$

where k_n is the linear momentum of particle n and V is an appropriate normalisation constant.

Within this approach, the matrix element becomes:

$$H_{if} = \frac{G}{V}\int \psi_p^* \psi_n e^{-i(\vec{k}_e + \vec{k}_\nu)\vec{r}} \, d^3\vec{r} \tag{1.18}$$

the exponential expression $e^{-i(\vec{k}_e + \vec{k}_\nu)\vec{r}}$ can be expanded and we obtain:

$$H_{if} = \frac{G}{V}\int \psi_p^* \psi_n \, d^3\vec{r} - i\frac{G}{V}(\vec{k}_e + \vec{k}_\nu)\int \psi_p^*(\vec{r})\psi_n(\vec{r}) \, d^3\vec{r} \, \dots \tag{1.19}$$

If the first term is non-vanishing, we talk about allowed transitions. The following terms gain relevance if the first term vanishes, as it would for example in the case of a transition between states of opposite parity. These transitions are called forbidden transitions.

Now let us return to Eq. 1.15 and discuss the consequences. It is possible to deduce the shape of the beta spectrum which played a key role in the development of the theory in its early days, since it provides a stringent way of testing it experimentally.

If the recoil energy of the nucleus is neglected, the energy available for the decay (E_0) is shared between the created electron and antineutrino.[2]

$$E_0 = E_e + E_\nu = Q + m_e c^2 \tag{1.20}$$

the density of states in Eq. 1.15 is composed of two factors, the density of ν states at the available energy $E_0 - E_e$ and the number of states $n_e(E_e)$, which represents the number of available electron states between $[E_e, E_e + dE_e]$.

The transition probability for decays in which the electron can have energies in the interval $[E_e, E_e + dE_e]$ is then:

$$d\lambda = \frac{2\pi}{\hbar} |H_{if}|^2 n_\nu (E_0 - E_e) n_e(E_e) dE_e \tag{1.21}$$

Neglecting spin, the density of states for the neutrino or the electron can be determined as:

$$n = \frac{V}{(2\pi)^3} 4\pi k^2 dk \tag{1.22}$$

which represents the number of states available for a free particle with momentum k in the momentum interval $[k, k + dk]$.

In Eq. 1.21 the density depends on the available energy, so we need to transform the dependence on k (Eq. 1.22) to a dependence on energy. For that we will use the first derivative of the relativistic relation:

$$E = \sqrt{p^2 c^2 + m^2 c^4} = \sqrt{(\hbar k)^2 c^2 + m^2 c^4} \tag{1.23}$$

$$\frac{dE'}{dk} = \hbar^2 c^2 \frac{k}{E} \tag{1.24}$$

Using (1.23) and expressing k from (1.24) we obtain:

$$d\lambda = \frac{|H_{if}|^2}{2\pi^3 \hbar^7 c^6} S_0(E_e) dE_e \tag{1.25}$$

$$S_0(E_e) = n_\nu(E_0 - E_e) n_e(E_e) \tag{1.26}$$

---

[2] In the case of a β^+ decay a positron and a neutrino are the relevant particles.

$$S_0(E_e) = \sqrt{(E_0 - E_e)^2 - m_\nu^2 c^4}(E_0 - E_e)\sqrt{E_e^2 - m_e^2 c^4}E_e \quad (1.27)$$

We see from this result that the energy dependence of the transition rate is dominated by the factors arising from the phase space of the leptons.

An additional correction factor was introduced by Fermi to take into account the effect the Coulomb field exerts on the created electron (or positron in the case of β^+ decay). This factor

$$F(Z_d, E_e) = |\frac{\psi_e(Z_d, 0)}{\psi_e(0, 0)}|^2 \quad (1.28)$$

is known as the Fermi function where $\psi_e(Z_d, \vec{r})$ is the electron wave function at position $\vec{r}$ in the Coulomb field of the daughter nucleus. Integrating Eq. (1.25) with the assumption that $m_\nu = 0$:

$$\lambda = \int d\lambda = \frac{m_e^5 c^4}{2\pi^3 \hbar^7}|H_{if}|^2 f(Z_d, E_0) \quad (1.29)$$

where λ represents the decay rate and

$$f(Z, E_0) = (\frac{1}{m_e c^2})^5 \int_{m_e c^2}^{E_0} F(Z_d, E_E)(E_0 - E_e)^2 \sqrt{E_e^2 - m_e^2 c^4}E_e dE \quad (1.30)$$

the f contains the effect of the available phase factor for the leptons and the impact of the Coulomb field on the electrons (or positrons in the β^+ case).

From these last equations we can see that

$$\frac{2\pi^3 \hbar^7}{m_e^5 c^4 |H_{if}|^2} = \frac{1}{\lambda}f(Z_d, E_0) \quad (1.31)$$

which is equivalent to the relation

$$\frac{2ln(2)2\pi^3 \hbar^7}{m_e^5 c^4 |H_{if}|^2} = f(Z_d, E_0)t_{1/2} \equiv ft_{1/2} \quad (1.32)$$

where we have used the relation $\lambda = \frac{1}{\tau} = \frac{2ln(2)}{t_{1/2}}$. In this last relation τ is the mean life, and $t_{1/2}$ is the half-life of the decay process.

Equation 1.32 is an important result, since it summarises the relation between the so-called $f t_{1/2}$ values, and the square of the matrix elements of the nuclear transitions $|H_{if}|^2$, where f is a calculated value (see (1.30)) and $t_{1/2}$ is the partial half-life of the beta transition determined experimentally.

> Transforming E to p_e in Eq. 1.27 shows that the shape of the beta spectrum is determined primarily by the statistical factor $p_e^2 p_\nu^2 = p_e^2 (Q - E_e)^2$, but the nuclear matrix element can contain an additional dependence on the electron(positron) and antineutrino(neutrino) momentum for forbidden transitions not discussed here. Precise beta shape spectroscopy, as required for precise deduction of the beta and antineutrino spectrum for fundamental applications might require additional factors and corrections. The details can be found in some recent publications [16, 17].

1.4 A Practical Example: The Beta Decay of ^{24}Na

As an example of a beta decay we show here the ^{24}Na $\rightarrow$ ^{24}Mg β^- decay. This decay is of interest in total absorption gamma spectroscopy, since it has an associated gamma-ray transition of relatively high energy, which is useful for the energy calibration of the detectors. The ^{24}Na decay also has the property that most of the beta intensity goes to one level, which makes it relatively simple to understand.

In Fig. 1.3 we see the information required to characterise the decay fully. The data presented are taken from the Evaluated Nuclear Structure Data File (ENSDF) [18]. Relevant quantities are: the Q value of the decay ($Q = 5515.45(8)$ keV) which, as mentioned earlier, is the energy difference from the ground state of the parent nucleus to the ground state of the daughter nucleus; the half-life of the decay ($t_{1/2} = 14.997(12)$ h), the spins and parities (J^π) of the states connected by the transitions, and the beta transition intensities, represented in the figure as the column I_{β^-}. Note that the beta transition intensity is by convention the probability of the decay to a particular level in the daughter nucleus multiplied by 100, see for example the beta transition to the state at 4122.9 keV excitation energy in the daughter nucleus in Fig. 1.3, where in 99.855% of the decays the transition proceeds to this state (0.99855 probability of decay to that level). All of these quantities ($Q, t_{1/2}, I_{\beta-i}$) are primary experimental quantities. There are also deduced values as the Log $f t_i$ values, which are the logarithms in base 10 of the $f t_i$ values discussed in the previous section for each level i. Here f is the function defined in formula (1.30) and $t_i = t_{1/2} * 100 / I_{\beta-i} = t_{1/2} / P_i$ is the partial half-life for the decay of the parent state to level i, defined as the half-life of the decay divided by the probability P_i of the beta

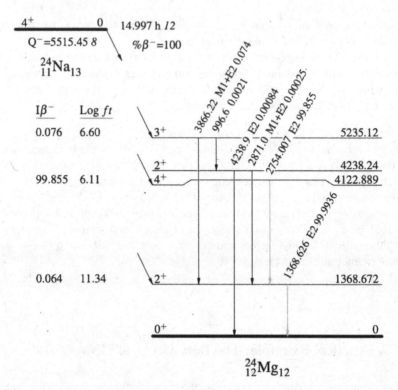

Fig. 1.3 Beta decay of the ^{24}Na source. The level at 4122.9 keV excitation energy receives most of the beta feeding and it de-excites predominantly to the ground state by a gamma cascade of two gamma rays with energies 2754.0 and 1368.6 keV. Level scheme downloaded from the ENSDF database

transition. Also in the figure we see the gamma transitions (and gamma cascades) that de-excite the populated levels and their absolute gamma intensities per 100 decays. ft values can be obtained easily using the logft tool provided by the National Nuclear Data Centre (NNDC) [19] if the required ingredients ($T_{1/2}$, P_i or $I_{\beta i}$, Q) are known.

The main topic of this book is how we determine the beta transition intensities (I_β) or the beta feedings with the total absorption technique, but before we introduce the technique we shall explain how these quantities can be measured. As we mentioned before because of the neutrinos (antineutrinos) emitted, the beta spectrum is a continuous spectrum independently of whether it is a β^+ or a β^- process. This makes it very difficult to disentangle the beta intensity going to a particular level by a direct measurement and analysis of the beta spectrum.

For that reason the beta intensities are determined indirectly relying on the detection of the gamma rays that are emitted following the beta decay process and assuming that we are able to determine somehow the feeding that goes directly to the ground state of the daughter nucleus. From Fig. 1.3 we see that any time an excited level is

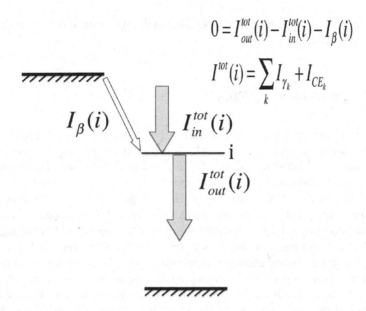

$$0 = I_{out}^{tot}(i) - I_{in}^{tot}(i) - I_\beta(i)$$

$$I^{tot}(i) = \sum_k I_{\gamma_k} + I_{CE_k}$$

Fig. 1.4 Intensity balance of transitions populating a level. The beta feeding is determined from the difference of the total gamma and conversion electron transitions de-exciting and feeding the level $I_\beta = I_{out}^{tot} - I_{in}^{tot}$

populated in the beta decay, a gamma cascade de-excites the level.[3] If we are able to construct the level scheme for the daughter nucleus populated in the beta decay, then it is possible to extract the beta feeding corresponding to the level from the difference in the total intensities of the transitions populating and de-exciting the level (see Fig. 1.4). It should be noted that the intensity populating a level is the sum of the intensity of electromagnetic transitions from upper levels and the incoming beta intensity ($I_{in}^{tot} + I_\beta$).

To extract the total intensity of the electromagnetic transitions populating and de-exciting a level, we need to detect the gamma rays emitted after the beta decay, place them in the level scheme and also determine the internal conversion coefficients of the transitions (see appendix), since the total intensity is relevant here. We also need to determine the fraction of the beta decays that goes from the ground state of the parent to the ground state of the daughter nucleus that is not accompanied by gamma emission.

Conventionally, in these experiments gamma detectors are used in combination with beta detectors. Among the available gamma detectors, Ge detectors are preferred because of their good energy resolution ($\Delta E/E \sim 0.2\%$) compared to other types of detector. The level scheme populated in the decay is constructed based on gamma-gamma coincidences, the intensity balance of the levels and the Ritz combination

---

[3] It should be noted that there are exceptions where a long-lived, isomeric state is populated in the decay. Then part of the cascade is delayed in time, and some of the transitions are converted.

principle. See Leo [20] and Knoll [21] for additional experimental details on gamma spectroscopy techniques.

1.5   The *Pandemonium* Effect

In some of the experiments mentioned earlier, problems may occur because of the limited gamma detection efficiency of Ge detectors. It means that one may fail to detect gamma rays emitted in the decay. This is a problem known as the *Pandemonium* effect and was introduced by Hardy and coworkers in Ref. [22]. In order to show the complexity faced in beta decay studies, Hardy et al. created an artificial nucleus *Pandemonium* using a statistical model and Monte Carlo (MC) techniques. This MC experiment was analysed using conventional analysis techniques and the undetected beta feeding was estimated. In their work they show that when beta feeding occurs to high lying levels in the daughter nucleus, many possible de-excitation paths can exist, and as a consequence weak gamma rays can be emitted. The decay of high-lying levels populated in the beta decay can also decay by the emission of high energy gamma rays. Weak gamma rays and gamma rays of high energy are difficult to detect if the gamma efficiency of the experimental setup is limited. As a consequence, if we build the decay level scheme based on a measurement with limited gamma efficiency, the feeding distribution determined will not correspond to the real situation (see Fig. 1.5). The solution to the problem is to increase the gamma detection efficiency dramatically in order to avoid missing any of the gamma rays. This can be achieved using the Total Absorption Gamma Spectroscopy technique (TAGS) discussed in the next chapter.

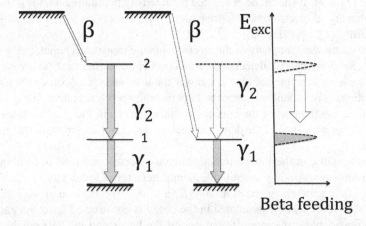

Fig. 1.5   Representation of the consequence of not detecting a gamma ray in the level scheme of the decay. The beta feeding is displaced to low-lying levels. This effect is known as the *Pandemonium* effect [22]

References

1. Becquerel, H.: Comptes Rendus. Académie des SCiences **122**, 420 (1896)
2. Becquerel, H.: Comptes Rendus, Académie des SCiences **122**, 501 (1896); Becquerel, H.: Comptes Rendus, Académie des SCiences **130**, 809 (1900)
3. Kaufmann, W.: Phys. Z. **2**, 602 (1901)
4. Rutherford, E.: Phil. Mag. **47**, 109 (1899)
5. Villard, P.U.: Comptes Rendus **130**, 1010 (1900); Villard, P.U.: Comptes Rendus **130**, 1178 (1900)
6. Rutherford, E.: Philos. Mag. Ser. 6 **27**(159), 488 (1914)
7. Geiger, H., Marsden, E.: Proc. R. Soc. Lond. A. **82**(557), 495 (1909)
8. Rutherford, E., Soddy, F.: Philos. Mag. IV **370–96**, 569–85 (1902)
9. Chadwick, J.: Proc. Roy. Soc. **136**, 692 (1932)
10. Kondev, F.G., Wang, M., Huang, W.J., Naimi, S., Audi, G.: Chin. Phys. C. **45**, 030001 (2021)
11. Chadwick, J.: Verh. Dtsch. Phys. Ges. **16**, 383 (1914)
12. Pauli, W.: Letter to nuclear physicists in Tuebingen, Germany (1930)
13. Fermi, E.: Il Nuovo Cimento **9**, 1 (1934)
14. Huang, W.J., et al.: Chinese Phys. C **45**, 030002 (2021); Huang, W.J., et al.: Chinese Phys. C **45**, 030003 (2021). https://www-nds.iaea.org/amdc/
15. Alvarez, L.W.: Phys. Rev. **52**, 134 (1937)
16. Huber, P.: Phys. Rev. C **84**, 024617 (2011); Mueller, T.A., et al.: Phys. Rev. C **83**, 054615 (2011)
17. Hayen, L., Kostensalo, J., Severijns, N., Suhonen, J.: Phys. Rev. C **100**, 054323 (2019)
18. ENSDF (2020). http://www.nndc.bnl.gov/ensdf
19. https://www.nndc.bnl.gov/logft/
20. Leo, W.R.: Techniques for Nuclear and Particle Physics Experiments. Springer
21. Knoll, G.F.: Radiation Detection and Measurement. Wiley
22. Hardy, J.C., et al.: Phys. Lett. B **71**, 307 (1977)

Chapter 2
Total Absorption Gamma Spectroscopy Basics

Abstract In this chapter the basic concepts of the total absorption gamma spectroscopy technique are introduced. Following the introduction, a brief description of the historical development of the technique will be presented. Later, the main challenges that need to be faced and an introduction to the analysis formalism are given.

2.1 Basic Ideas

As mentioned in the previous chapter, the intensity of the beta feeding to levels in the daughter nucleus is frequently determined indirectly through the detection of gamma rays. In this context high resolution detectors (Ge detectors) are very useful, since they allow us to carry out detailed spectroscopy of the states populated in the daughter nucleus, including the precise energies of the levels, their decay patterns and their quantum characteristics. In other words, they allow us to establish beta decay schemes. However even at the technological level available today, the most commonly available Ge detection systems for beta decay studies have serious limitations in efficiency.

Total absorption gamma spectroscopy, as the name indicates, aims to detect all gamma radiation emitted following beta decay. To reach that goal we need another type of detector, with much higher detection efficiency that acts like a calorimeter by covering the source in a 4π geometry. Figure 2.1 shows an ideal device, that can be constructed by surrounding the source with a shell of detector material that should allow for the full detection of the gamma cascades that follow the beta decay. Such a detector can be constructed with scintillator materials. Figure 2.1 shows also how a simple decay that proceeds only to one level, is seen by ideal detectors. The emitted beta particle spectrum detected by a Si detector is presented in the upper panel, where we can see its continuous character. The gamma radiation emitted after the beta decay, as detected by a Ge detector system is presented in the middle panel, where the two peaks from the gamma rays that de-excite the populated level are seen. The total absorption gamma spectrum detected by the total absorption gamma spectrometer (TAGS) (lower panel) shows one peak, representing the populated level

A. Algora et al., *Total Absorption Technique for Nuclear Structure and Applications*, SpringerBriefs in Physics, https://doi.org/10.1007/978-3-031-58864-8_2

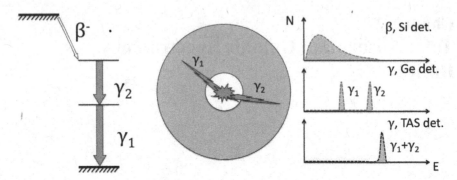

Fig. 2.1 Left side: simplified beta decay where only one level is populated in the daughter nucleus that de-excites by a gamma cascade of two gamma rays. Centre: idealised representation of a TAGS detector detecting one decay event. Right side: the decay seen by different detectors under ideal conditions

in this ideal example. From this figure we can see why the total absorption technique is different. In ideal conditions the spectrum shows directly the populated levels, and their intensities in the spectrum would be proportional to the feeding distribution. So, with a total absorption spectrometer we do not see the individual gamma rays emitted in the decay, we see the sum of the energy of the gamma cascades that follows the beta decay.

Figure 2.1 represents schematically an ideal situation for several reasons. On the one hand, the presented case is very simple, since only one level is populated in the beta decay. On the other hand, the ideal Ge detector system is able to detect the two emitted gamma rays, so there is no *Pandemonium* effect. Finally the total absorption spectrum only sums the gamma rays (there is no penetration of beta particles) and hence the detector acts as a 100% efficient gamma summing device. It is clear that this example does not match real situations since many levels can be populated in a typical beta decay. As a result many gamma rays may not be observed, because an array of Ge detectors is not sensitive enough, and there may also be penetration of the beta particles or other type of accompanying radiation in the total absorption spectrometer.

A real total absorption spectrometer can not be constructed as a 4π closed shell of detector material, since we need to position sources in the centre of the detector for the measurements. The scintillator material, from which the detector is constructed, has to be canned, to protect it from the ambient light. This may also be necessary if the material is hygroscopic. In this type of measurements it is common to use ancillary detectors,[1] such as beta detectors or X-ray detectors and/or a tape transport system

---

[1] Ancillary detectors are used to study the beta decay in coincidence mode. For example a Si detector can be used to tag events in which a β^+ or a β^- particle is emitted. An X-ray detector can be employed to detect the X-rays that are emitted in the electron capture (EC) process and tag those events. Coincidence conditions contribute in general to providing a cleaner spectrum free from ambient background.

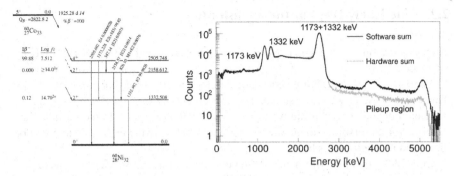

Fig. 2.2 Left panel: decay of ^{60}Co as included in the ENSDF data base. Right panel: total absorption spectrum from the ^{60}Co decay taken by the segmented detector DTAS at Jväskylä IGISOL IV facility [1] where the dominant sum peak (1173 + 1332 keV) from the cascade is clearly seen as well as the single peaks (1173 and 1332 keV) arising from the incomplete sum. Single peaks from the cascade arise because the total absorption spectrometer does not have 100% efficiency

to complement the total absorption setup. The former helps to identify the decay of interest and the latter is necessary to carry radioactive material to the counting position within the detector and remove sources that have decayed. All these constraints make it impossible to construct a 100% efficient summing device because we need to make holes and add dead material to the setup. For that reason, a real total absorption spectrum will look like the one shown in Fig. 2.2 where we do not only see sum peaks corresponding to the levels fed in the decay, but we also see escape peaks when the gamma cascade is not fully absorbed as well as continuous background. The analysis of such an spectrum will require solving the following equation that will be discussed in great detail in the chapters that follow namely:

$$d_i = \sum_{j=0}^{j_{max}} R_{ij}(B) f_j + C_i \tag{2.1}$$

where d_i is the content of bin i in the measured TAGS spectrum, R_{ij} is the response matrix of the TAGS setup and represents the probability that a decay that feeds level j in the level scheme of the daughter nucleus gives a count in bin i of the TAGS spectrum, f_j is the β feeding to the level j (our goal). The index j in the sum runs over the levels populated in the daughter nucleus in the β decay. C_i represents the contribution of different distortions and contaminations to the real spectrum.

2.2 The Early Days of the Technique

Total absorption gamma spectroscopy was first introduced at CERN-ISOLDE by Duke et al. in 1970 [2]. The first spectrometer was constructed from two $\emptyset = 150\,mm \times l = 100\,mm$ NaI detectors placed face to face at a separation of 15mm. This detector was used for systematic measurements of the beta decay of neutron-deficient nuclei between Iridium ($Z = 77$) and Radon ($Z = 86$). In their article they describe for the first time many of the challenges faced by the technique that will be discussed in detail in this book.

In 1973 the first measurements of the beta decay of neutron-rich nuclei were performed by Johansen et al. [3] at the OSIRIS facility in Studsvik (Sweden) using this technique. In their measurements the same detector was used with one modification: they included a ring-shaped beta detector placed around the region that separated the NaI detectors. The main idea was to use coincidences with beta particles, that have already left the volume between the NaI detectors and consequently did not generate summing with the gamma cascades in the analysis of the TAGS spectrum. The motivation behind this idea was to simplify the calculation of the response function, that represented a serious challenge for the technique. As in the article of Duke et al., the experimental difficulties faced in the measurements were discussed in detail. Both Duke et al. and Johansen et al. describe the technique as *incomplete total absorption*, due to the limited size of the detectors used and devote considerable effort to the analysis of the spectra. In Ref. [3] Monte Carlo techniques were used to calculate the response function of the detectors. More details on the early efforts can be found in Erdal and Rudstam [4].

The article of Duke et al. also introduces the beta strength as the inverse of the ft values mentioned in (1.32) in a statistical sense. The level density at high excitation energy in the daughter nucleus makes it necessary to use the product of the level density and an average transition probability for the description of the beta decay process. This product is called the strength function [2, 5]:

$$S_\beta(E) \equiv \frac{I_\beta(E)}{f(Z_d, Q - E)T_{1/2}} = \frac{1}{f(Z_d, Q - E)t_{1/2}} = D^{-1} \sum_i \overline{B_i(E)}\rho_i(E) \qquad (2.2)$$

where $I_\beta(E)$ is the absolute beta feeding to levels at the excitation energy E, $Q - E$ is the endpoint of the beta transition or in other words the energy available in the beta transition to the level at excitation energy E, and f is the statistical rate function. Please note that $I_\beta(E)$ represents the feeding (normalized to 1), but we have used a different letter to avoid confusion with the f statistical rate function. The last part of the equation represents the theoretical definition of the strength, where $\overline{B_i(E)}$ is the averaged reduced beta-transition probability, $\rho_i(E)$ is the level density in the daughter nucleus and D is a constant. B is related to the nuclear matrix element (introduced in the earlier chapter) that connects the initial and final states. Actually, the first total absorption measurements were motivated to study systematically the dependence of this quantity on the excitation energy of the daughter nucleus. This motivation is still one of the major goals of present-day total absorption studies.

Another development occurred in the Soviet Union in the 1980s, where Bykov et al. [6] used a different geometrical arrangement for the total absorption detector. Their spectrometer consisted of a NaI(Tl) crystal of cylindrical shape with $\varnothing = 200$ mm $\times l = 200$ mm with a $\varnothing = 40$ mm $\times l = 100$ mm axial well that was complemented with an additional NaI(Tl) detector to close the well. This detector was used for systematic studies of neutron-deficient nuclei. Alkhazov et al. [7], provides an example, showing the results of the study of the beta decay of ^{145}Gd. This study was triggered by the debate between Hardy and Firestone about the completeness of level schemes populated in the beta decay studied using conventional Ge spectroscopy [8–13]. Alkhazov et al. concluded that it is necessary to apply the total absorption technique to avoid the *Pandemonium* effect.

In the 1990s Greenwood et al. [14] used a NaI(Tl) spectrometer of similar shape with dimensions $\varnothing = 254$ mm $\times l = 305$ mm and a $\varnothing = 51$ mm $\times l = 203$ mm axial well. This detector was employed for systematic studies of the beta decay of fission products at the Idaho National Engineering Laboratory (INEL) ISOL facility. The crystal was larger than the systems used previously [2, 6] and provided a better approximation to the ideal detector. This detector produced beta decay data that are still considered useful today in applications to reactor decay heat and the calculations of the antineutrino spectrum as will be discussed later.

An even larger detector was built by Nitschke et al. [15] and commissioned at the Lawrence Berkely Laboratory (LBL). However since the online apparatus for the Super HILAC Isotope Separator (OASIS) facility was closed, the detector was moved to GSI. At this laboratory it was used at the Mass Separator facility for measurements of decays in the ^{100}Sn and ^{146}Gd regions. This detector has a cylindrical shape $\varnothing = 356$ mm $\times l = 356$ mm with a $\varnothing = 51$ mm $\times l = 203$ mm axial well filled with a NaI(Tl) plug detector. The detector was combined with ancillary detectors, a Ge X-ray detector and Si detectors, placed in the well for disentangling electron capture and beta plus processes and also included a novel gain stabilisation system.

In the early 2000s, an even larger detector was installed at CERN-ISOLDE. It was named *Lucrecia* and is made of a single crystal of NaI(Tl) with a cylindrical shape of dimensions $\varnothing = 380$ mm $\times l = 380$ mm. At the time it was built, it was considered to be the largest NaI(Tl) mono-crystal detector (see Fig. 2.3) in the World. *Lucrecia* has a cylindrical hole of $\varnothing = 75$ mm diameter that goes through the crystal perpendicular to its symmetry axis. This detector has been used in nuclear structure studies related to the determination of the shape of the ground state of the decaying nucleus as will be discussed later. The reader will find a recent review of studies performed with *Lucrecia* in Rubio et al. [16].

All these detectors paved the way for a new generation of segmented detectors such as *Rocinante* [17], DTAS [18], SUN [19] and MTAS [20] that will be discussed in more detail in another chapter. The segmentation of these new detectors can provide additional information on the gamma multiplicity of the gamma cascades that can be useful in the complex analysis.

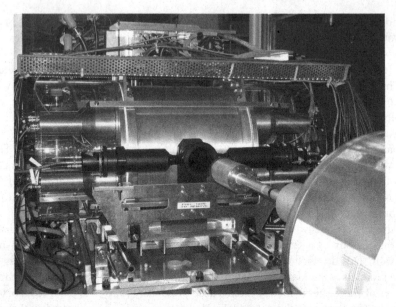

Fig. 2.3 Photo of the *Lucrecia* spectrometer installed at CERN. In the photo the ancillary detectors typically used can also be seen. The black structure in the centre of the detector is a plastic detector for the detection of beta particles and on the right a Ge telescope detector detector can be seen. During the measurements both the plastic detector and the Ge telescope are placed inside *Lucrecia*

2.3 Challenges

We will now present the major challenges to the use of TAGs and the basic ideas behind the analysis of data acquired using it. A detailed technical description of the different algorithms used for the analysis will be presented in the next chapter.

The major challenge when constructing a spectrometer is to maximise the efficiency of the setup. This can be limited by the availability of detector materials and the construction cost. In the early days of the technique all spectrometers were constructed from NaI(Tl), which is a material that provides reasonably good efficiency for moderate dimensions and cost, and a reasonable energy resolution of $\Delta E/E \sim 6-8\%$ at 662 keV. If a faster detector is needed, BaF_2 provides a suitable alternative [17]. This last material has additional advantages: one is that it has a lower neutron capture cross-section than NaI(Tl), which can be relevant when studying beta decays involving beta delayed neutron emission, since the neutron capture in the detector material can result in contamination of the total absorption spectrum. The other advantage is that the generation of scintillator light in the detector material is less non-linear than the NaI(Tl) case. Unfortunately, the material has poorer energy resolution than NaI ($\Delta E/E \sim 12-14\%$). In the last decade new scintillator materials have become commercially available and bigger crystals can be grown. Materials like $LaBr_3$:Ce or $CeBr_3$ may have energy resolutions of the order of $\sim3\%$ and are faster than NaI(Tl). These materials, like BaF_2 may also have internal contamination, but this is a minor problem if coincidence techniques with ancillary detectors are used in the analysis. In some sense it can be considered an advantage. The internal contamination can be used to monitor the gain-shifts of the system and to correct it as has been done in the case of BaF_2 [17]. The high cost of the new scintillator materials has made it impossible until now to construct a total absorption spectrometer made fully of such materials, but hybrid geometries have been considered, where detectors made of $LaBr_3$:Ce can be added to an existing NaI(Tl) detector to improve the energy resolution of the individual modules [18].

The relevance of maximising the efficiency of total absorption setups cannot be underestimated. The closer the efficiency of the setup to the ideal of 100% efficiency, the lower will be the dependence of the results of the analysis on the response of the detector to different de-excitation decay paths. But maximising the efficiency also has a counter effect. The detector will also be more efficient in detecting undesired radiation, thus increasing the background in the measurements. Depending on the setup, a simple lead layer or more sophisticated multilayer shieldings have been used to reduce the effect of gamma and neutron radiation on the detector. For example, in the *Lucrecia* setup at ISOLDE [16] the room background counting rate can be reduced from 3000 counts/s with open shielding to 1000 counts/s with closed shielding. In this case the shielding is composed of the several layers: first polyethylene to reduce the impact of neutrons coming from the production reactions and the hall, secondly layers of lead, copper and tin to reduce the impact of ambient gamma-rays and X-rays.

The use of coincidences is also useful in reducing the contribution of the background radiation, including the possible internal contamination of the detector crystals as was mentioned earlier. But even in that case the use of passive shielding is useful to improve the signal to background ratio during the experiments and to reduce the unnecessary loading of the data acquisition systems with undesired events.

Coincidences with ancillary detectors are also useful to select the beta-decay process of interest. For example, when studying proton-rich nuclei [15, 16] one may be interested in separating the contribution of the β^+ decay process from that from electron-capture (EC). As mentioned in Chap. 1, the EC process is accompanied by the emission of characteristic X-rays in the daughter nucleus, so it can provide element selective data when TAGS data are collected in coincidence with an X-ray detector. However the separation of processes based on gating with the X-rays should be done with care, since de-exciting level transitions in the daughter can also induce X-rays through the internal conversion process (IC) [21]. If some transitions in the daughter nucleus are highly converted the X-ray gated TAS spectrum cannot be considered to be due solely to EC, and it is probably contaminated by the β^+ process, since it is not possible to distinguish if the X-ray used for tagging the event comes from a conversion process (after β^+ or EC) or from the EC process.

β^+ and β^- processes can be selected using coincidences with an ancillary beta detector. These cases are not as clean as the X-ray gated EC spectrum, since we do not have element selectivity from the direct measurement of the β spectrum. These gated spectra can also be contaminated by daughter, grand-daughter, etc. decays, that require separate measurements to evaluate their contribution.

In these measurements it is very relevant to have a proper assessment of the beta efficiency of the detector, since the β efficiency will enter in the calculation of the response function of the total absorption setup. This is important in the analysis of both β^+ and β^- decays. It should be noted that due to electronic thresholds in the beta detectors and the continuous nature of the beta spectrum there is a strong dependence of the beta efficiency on the end-point energy that needs to be determined, typically by a combination of Monte Carlo simulations and benchmark measurements [22, 23].

Another challenge is how to handle the electronic pileup of signals. The overlap of two signals in the detection system can generate a higher energy signal that can be mistaken for a real one. This distortion has to be estimated in total absorption experiments, since it can affect regions of the spectrum where there is a reduced number of counts (for example at excitation energies near the Q value of the decay). In such a situation, counts in the TAS spectrum arising from electronic pileup can lead to incorrectly deduced beta feedings at high excitation energy, which will subsequently lead to a large systematic error in the determination of the beta strength. This problem was addressed in [24] for the total absorption technique and studied in great detail. The idea is to calculate the pileup using a true signal pulse, which is measured with an oscilloscope, or a digital system, during the experiment. In the algorithm all possible overlaps of two signals within the acquisition gate of the measuring system are calculated. The signals are generated using the measured mean electronic pulse shape with amplitudes generated from a Monte Carlo sampling of the true measured total absorption spectrum. In [24] the algorithm was developed for a peak sensing analogue to digital converter (ADC), but later it has been extended to other kinds of ADC. This correction is even more complex for segmented systems, where in addition to the conventional pileup when we have the overlap of signals in one module of the multidetector system, we can have the effect of different decays detected by the different modules. This contribution is called "summing" and it is more relevant than the pileup in the simple modules. An algorithm to calculate this correction can be found in [25].

As mentioned in this section, identifying the possible contaminants of the total absorption spectrum is a very important task, since the analysis of the data is based on the full spectrum and not on the identification and integration of individual gamma peaks as is the case in higher-resolution spectroscopy with Ge detectors. This problem has two aspects: first the determination of all possible contaminants, and secondly the quantification of their contribution to the spectrum to be analysed. Depending on the case several strategies can be followed. For example, the contribution of the pileup can be normalised to regions of the spectrum, where no decay counts are expected, beyond the largest decay Q value. In the case of measurements in singles mode (not in coincidence with ancillary detectors), a similar strategy can be followed for the normalisation of the background. So, we should look for regions of the spectrum that clearly are related to the contamination in order to determine their normalisation factors. In the case of daughter, grand-daughter decay, etc. contamination, we can look for characteristic peaks, or for regions where this contamination is dominant and use that information to determine their contribution to the spectrum.

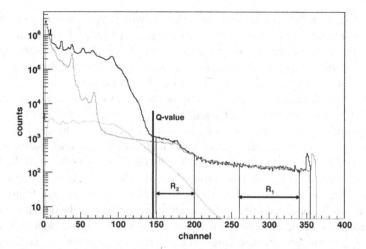

Fig. 2.4 The figure shows a typical total absorption spectrum with contributions from various effects. The case of ^{104}Tc decay is shown [26]. The measurement was performed in singles mode. The background is normalised to region R_1 and the contribution of the pileup is normalised to region R_2 using the relevant areas of the spectrum

$$d_i = d_i^{clean} + C_i = d_i^{clean} + a * c_i^a + b * c_i^b + \cdots \qquad (2.3)$$

Equation 2.3 represents the various contributions to channel i of the total absorption spectrum, where d_i^{clean} represents the total absorption data of the decay of interest and C_i the contribution of all the contaminants. c_i^a, c_i^b, etc. represent the different contributions of the contaminants to channel i and a, b, etc. their normalisation factors. See Fig. 2.4 for an example.

2.4 Introduction to the Analysis

As mentioned earlier in this chapter, the analysis of the spectra recorded in a total absorption measurement requires the solution of the following equation:

$$d_i = \sum_{j=0}^{j_{max}} R_{ij}(B) f_j + C_i \qquad (2.4)$$

where d_i is the content of bin i in the measured TAGS spectrum, R_{ij} is the response matrix of the TAGS setup and represents the probability that a decay that feeds level j in the level scheme of the daughter nucleus gives a count in bin i of the TAGS spectrum and f_j is the β feeding to the level j, which is our goal. In line with the discussion of the earlier section we have added C_i, which represents the contribution of all contaminants to the channel i to the equation. The index j in the sum covers the levels populated in the daughter nucleus in the β decay. d_i can represent a singles spectrum, where only the TAGS detector is used, or a spectrum where signals in the TAGS are in coincidence with signals from ancillary detectors.

The solution of Eq. 2.4 in order to determine the values of f_j-s, requires a number of preparatory steps. Firstly, as discussed in the earlier section, we need to quantify the contribution of the contaminants C_i to the measured data. The other necessary condition for solving the equation is the determination of the response matrix $R_{ij}(B)$. The response R depends on the setup and on the de-excitation pattern of the levels population in the daughter nucleus through the so-called branching ratio matrix B. In practice B means the knowledge of the de-excitation pattern (to low-lying levels) of every possible populated level in the daughter nucleus. Thus, the determination of $R(B)$ means that we know how our setup responds to all possible populated levels in the daughter nucleus. In most cases all the necessary information that defines B is not available. The reason is evident: in most cases the level scheme populated in the decay is poorly known or at least incomplete.

To overcome this difficulty assumptions must be made. To calculate $R(B)$, first we define a possible B branching ratio matrix, which can later be modified during the analysis if necessary. For that we can use the information available in evaluated nuclear data bases [29], but we need to look at the available information critically. We establish a "cut" or threshold excitation energy, which represents the energy up to which the known level scheme can be considered complete in terms of the number of levels. Up to that cut energy the levels available in evaluated data bases and their decay pattern are used. This is based on the assumption, that probably the low lying levels and their de-excitation can be well defined in studies using Ge detectors. This information is then complemented with levels and branching ratios calculated using the statistical model. To do that the energy range between the cut energy and the decay Q value is divided in energy bins, typically of 40 keV width in our analyses. The probability of finding a level with a particular spin and parity in an energy bin, and how these levels de-excite to levels in the calculated lower-lying energy bins, or to the better defined levels in the lower part of the level scheme is calculated using level density functions and gamma strength functions of E1, M1 and E2 character. In practical terms we determine a possible realisation of the level scheme of the daughter nucleus (B) compatible with the best available knowledge. Once the branching ratio matrix is defined the response matrix ($R_{ij}(B)$) of the setup to a particular level defined in B is calculated using Monte Carlo techniques. Figure 2.5 shows a schematic picture of the division of the daughter level scheme. Figure 2.6 shows the response matrix used in the analysis of ^{152}Yb decay (see Refs. [27, 28] for more details.)

Before expanding further, we shall look again at what was meant earlier by a complete level scheme up to a certain excitation energy. This is not a trivial matter, since we need to know all of the energies, spins and parities of these levels. In practice this means that we have to assess if spins and parities are properly assigned to the levels and if the number of levels observed is in agreement with a model.

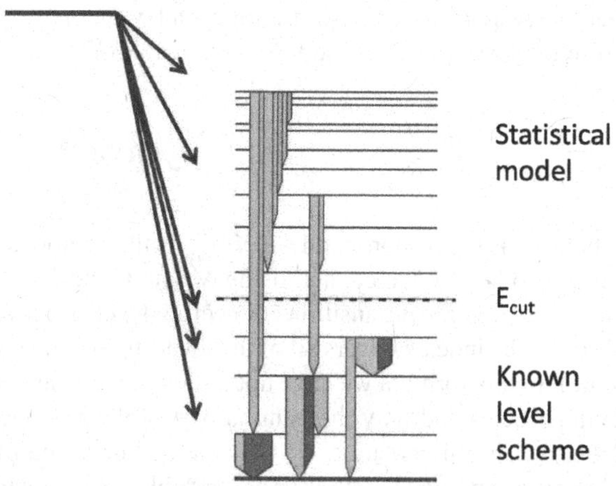

Fig. 2.5 Typical division of a level scheme used in the analysis of a TAGS spectrum. Split lines in the known part represent the competition between gamma emission and internal conversion process

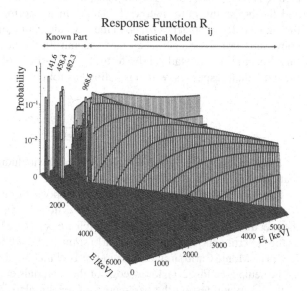

Fig. 2.6 The figure shows the response function used in the analysis of ^{152}Yb decay [27]. The E axis represents the distribution of the energy deposited in the total absorption detector by gamma cascades. The E_x axis represents the excitation energy of the levels in the daughter nucleus

The Reference Input Parameter Library (RIPL-3) database [31] can provide such information. For example we can see that the available level scheme for ^{104}Ru can be considered complete up to level 34 (from 70 known levels) at an excitation energy of 2.6189 MeV. This database takes data from the Evaluated Nuclear Structure Data File (ENSDF) [29] and NUBASE [30] and compares the existing data with predictions from the constant temperature model that gives the level density function. For more details see [31].

Once the branching ratio B or the level scheme is defined, one can calculate the response function of the setup for that particular level scheme. Since we do not have means to measure all possible mono-energetic gamma responses that are needed, this can only be calculated using Monte Carlo techniques. Then, based on the individual gamma responses, the response for each level can be determined recursively starting from the lowest level in the following way [32]:

$$R_j = \sum_{k=0}^{j-1} b_{jk} g_{jk} \otimes R_k \qquad (2.5)$$

where R_j is the response to level j, g_{jk} is the response of the gamma transition from level j to level k (calculated using Monte Carlo simulations), b_{jk} is the branching ratio for the gamma transition connecting level j to level k, and R_k is the response to level k. The index k covers all of the levels below the level j (0 represents the ground state). In this formula we have not included the convolution with the response of the beta particles and only the gamma part of the response is presented for simplicity. Note that in this notation R_j is a vector that contains as elements the R_{ij} matrix elements mentioned above for all possible i-s (or channels) in the TAGS spectrum and the branching ratio matrix (B) enters in the formula of the response matrix through the decay branches b_{jk}-s. In the real calculation of the responses the internal conversion process is also taken into account. In the next chapter we shall look at this aspect in more detail. In the formula the operator $\otimes$ means the convolution of the individual responses, which is defined as follows:

$$[p \otimes q]_i = \sum_{k=0}^{i} p_{i-k} q_k \qquad (2.6)$$

The responses in (2.5) are normalised to one, by including a fictional zero channel, that represents the non-interaction probability.

Prior to calculating the different gamma responses of the spectrometer, the Monte Carlo simulations have to be validated. This is done by reproducing measurements of the spectra from radioactive sources, which are performed under the same conditions as the experiments. For the Monte Carlo simulations the GEANT4 code is now used [33]. The Monte Carlo simulations require a detailed implementation of the geometry of the setup (see Fig. 2.7), knowledge of the materials employed in the construction of the setup and testing the parameters of the simulations to find the best tracking options and physics models that reproduce the spectra from the measured sources.

After identifying the different contaminants and calculating the response function, we can solve Eq. 2.4. There are several algorithms available for solving the equation [34]. This will be discussed in detail in the next chapter. One key question is to decide what is a good solution to the problem. The algorithms that we use conventionally are implemented iteratively, and at each step, the quality of the reproduction of the spectrum can be assessed by comparing the d_i (measured spectrum) with $\sum_{j=0}^{jmax} R_{ij}(B) f_j^n + C_i$ (calculated or reconstructed spectrum), where the n index represents the nth iteration or the final result. So, we compare the reconstructed spectrum which is obtained by multiplying the response function of the detector with the feeding distribution obtained from the analysis, at step n. Depending on the case we add the contaminants to avoid subtractions. For the comparison a χ^2 probe can

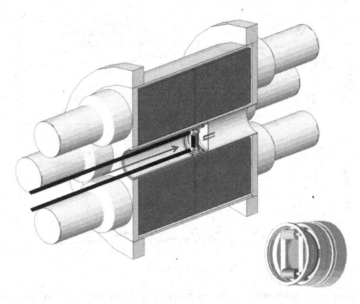

Fig. 2.7 Geometry implemented in GEANT4 of the total absorption spectrometer *Rocinante*. This spectrometer was employed in experimental campaigns at the IGISOL IV facility in Jyväskylä [17]. In the inset, the endcap of the tube with the Si beta detector is shown (not to scale). The thick black lines represent the magnetic tape from the tape system and the blue line represents the direction of the beam. Copyright (2010) by the American Physical Society

be used. As part of the analysis procedure, the assumptions used for generating B, such as the cut energy, the assigned spins and parities of the levels, or the parameters of the statistical model, can be modified. This is done within the accepted range of physical parameters. So in practical terms, in the analysis the following steps are followed until the best description of the data is obtained: (a) define a branching ratio matrix B, (b) calculate the corresponding response matrix $R_{ij}(B)$, and (c) solve the corresponding Eq. 2.4 using an appropriate algorithm and (d) compare the generated spectrum after the analysis $(R(B)f + C)$ with the experimental spectrum d (see for example Fig. 2.8).

In this chapter we have described up until now how the analysis is performed by our group. This is based on solving Eq. 2.4 for the f_j (that represents the problem) by an algorithm once the elements of the equation are well defined. This way of addressing the problem can be called the formal way. But this is not the only way of solving the problem. Other alternative methods have been used. For example one practical option is to start from what is known from Ge detector measurements and compare Monte Carlo simulations based on the known level scheme with the measured spectrum. If the result is not appropriate the feeding distribution can be changed and/or additional levels can be added by hand until a proper description of the experimental spectrum is achieved. To apply this method a Monte Carlo code with an appropriate event generator is needed. By visual comparison of the Monte Carlo

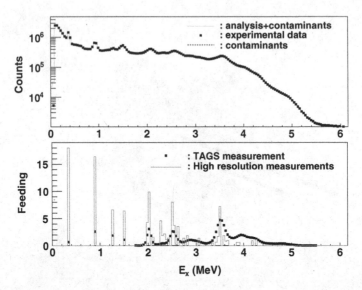

Fig. 2.8 Upper panel: comparison of the measured total absorption spectrum of the decay of [104]Tc with the spectrum generated based on the analysis. The generated spectrum is obtained by multiplying the response matrix with the feeding distribution. The contribution of the contaminants is also shown. The lower panel shows the comparison of the feeding distribution obtained from the TAGS analysis with the previous Ge detector studies [35]. Copyright (2010) by the American Physical Society

result with the measured spectrum, the feeding distribution can be optimised and excitation regions in the daughter nucleus where levels are missing can be identified. As part of this method additional "pseudo" levels can be added. The tricky part here is how to define the de-excitation branchings of the newly added levels, or in other words, how they are connected to the lower lying levels. This method of analysis was used in the earlier days of the technique (see for example [14, 36]). Some results obtained with this technique by Greenwood et al. [14] have been compared with the results obtained using our method of analysis. They provided very similar results [37, 38].

The new generation of segmented detectors such as DTAS, MTAS, SUN [18–20] provides additional information for testing the assumptions employed in the analysis. The fold, the number of detectors that fires in a TAGS event, can be used to test the analysis assumptions, since this number is directly related to the gamma-cascade multiplicity as a function of excitation energy and ultimately to the de-excitation branching ratio matrix B. A lack of knowledge of the matrix B is the largest source of uncertainty in TAGS analysis and this can be greatly improved with the use of segmentation. Our present approach to the iterative procedure for updating B described earlier in this Section, is to include in step (d) the comparison to fold-gated TAGS spectra and single module spectra. Reconstructed fold-gated spectra are obtained by MC simulations using an appropriate event generator since it is not possible to define a fold-gated response in a manner similar to Eq. 2.1. A different approach has been taken by the ORNL group to analyse MTAS data [39, 40]. They use the coincidence between one module and the sum of all the modules to define total energy gated single detector spectra that can be fitted by the sum of a number of de-excitation cascades, usually taken from Ge detector spectroscopy and supplemented when necessary with "pseudo" levels and postulated branching ratios. They are then modified iteratively until the best reproduction is achieved. Another approach is used at NSCL to analyse SUN data. [41]. They start from the same total energy gated single detector spectra but apply the so called Oslo-method [42] to obtain the branching ratio matrix for a subset of levels. The TAGS analysis is not performed with this B but uses the "pseudo" level approach including in the fit the total absorption spectrum and the spectrum of detector multiplicities [43]. More recently the NSCL group has been using an approach based on the DICEBOX statistical code [44] to calculate the branching ratio matrix.

2.5 Practicalities

As discussed in earlier sections the Monte Carlo technique plays a key role in the total absorption spectrum analysis. For that reason it is worth discussing briefly how the Monte Carlo simulations are performed and how they are handled when they are compared with experiment. Such comparisons are needed in two phases of the work. Firstly when the Monte Carlo simulations are optimised and validated in comparison with spectra recorded with sources, where the decay is well known, measured under the same conditions as in the experiment. Another necessary step, mentioned earlier, is the comparison of $\sum_{j=0}^{j_{max}} R_{ij}(B) f_j^n + C_i$ at the end of each iteration n of the analysis procedure with the measured spectrum d_i, since the Monte Carlo simulations are required to calculate the ingredients of R_{ij}.

In Monte Carlo simulations the energy deposited in the active volume of the detector is collected after the generation of each event. Typically the results of these simulations have no energy resolution if no instrumental width has been set in the simulations. In that case, when comparing the simulations with experiment, we need to alter the collected Monte Carlo energy spectrum so that it has the true experimental resolution. For that purpose we need to determine the energy dependence of the experimental resolution of our detector. In the procedure, peaks, escape peaks or single peaks of known energy from sources are employed. After fitting those transitions typically with a combination of a Gaussian and a step function, the energy dependence of the resolution of the experimental data $\sigma_{exp}^2(E)$ is determined:

$$\sigma_{exp}^2(E) = a_{exp} E + b_{exp} E^2 \tag{2.7}$$

This dependence can be used to widen the results of the "zero" resolution Monte Carlo simulations.

In this context it is worth mentioning that one might need different strategies for the widening of the Monte Carlo simulations depending on the detector material. In the total absorption spectrometer one collects the light generated in the scintillator material by the interactions of the gamma-rays or by the penetration of the betas particles in the detector. So one might wonder if collecting the deposited energy in the Monte Carlo simulations is sufficient for the description of the experimental data or a more complex procedure is needed. It turns out that this depends on the detector material. In the case of BaF_2 the conventional procedure of collecting the energy deposited in the detector is sufficient, since the light generated in this material is proportional to the deposited energy. NaI(Tl) represents a more challenging case because of the non-proportionality of the generated light. For that reason, when working with NaI(Tl) detectors, at the Monte Carlo level we generate and collect "light"[2] in the detectors from the simulated interactions in the detector material. This procedure was described in [32]. We use a light generation function at the step level of the Monte Carlo simulations. Because of the light generation implemented in the simulations, the Monte Carlo spectra for NaI(Tl), have a non ideal (or zero) width:

$$\sigma_{MC}^2(E) = a_{MC} E + b_{MC} E^2 \qquad (2.8)$$

and one needs to extract the dependence of the resolution of the Monte Carlo simulations on the energy (see Eq. 2.8). Prior to the MC versus experiment comparison one needs to widen the Monte Carlo simulations with the following dependence (see Eq. 2.9):

$$\sigma_{inst}^2(E) = \sigma_{exp}^2(E) - \sigma_{MC}^2(E) = a_{inst} E + b_{inst} E^2 \qquad (2.9)$$

and recalibrate the experiment to the Monte Carlo simulations for the comparison. When light is incorporated in the simulations an energy calibration of the Monte Carlo simulations is also needed for the comparison.

The non-linear response of the generated light in NaI(Tl) was already known from earlier studies [45], but its relevance in total absorption spectroscopy is first discussed in detail in [32]. One consequence of the non-linear response is that if the energy calibration of the experimental spectrum is obtained using single or escape peaks, the energy of sum peaks of two coincident gamma rays are displaced in the spectrum by approximately 30 keV, the sum of three coincident gammas are displaced by about 60 keV, etc. with respect to the sum of the energies of the single peaks. If this effect is not taken into account in the simulations, the reproduction of the experimental spectrum cannot be done properly in the analysis. As mentioned earlier this can be solved by implementing in the Monte Carlo simulations the light generated by the secondary particles in the steps of the simulation. For that the function L_e/E is used, which represents the ratio of the light yield of a fully absorbed electron of energy E in the NaI(Tl) material:

---

[2] Here we do not mean the generation and tracking of optical photons as is possible with Geant4 for example, but the emulation of the non-proportionality effect in the response.

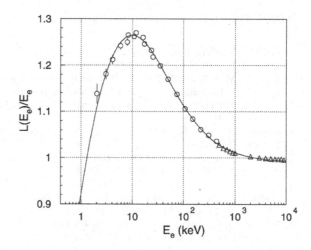

Fig. 2.9 NaI(Tl) relative light yield values for fully stopped electrons as a function of their energy. For more details see [32]

$$\frac{L}{E_e} = \frac{a_1(1 - e^{(-a_2 E_e)} + a_3 E_e + a_4 E_e^2}{a_5 + a_6 E_e + a_7 E_e^2} \qquad (2.10)$$

$$\{a_i\} = \{1.6(2), 0.058(8), 0.580(4), 0.00490(2), 0.25(2), 0.479(4), 0.00494(2)\} \qquad (2.11)$$

where the parameters $\{a_i\}$ are taken from [32].

This function is presented in Fig. 2.9. The work of [32] shows that the light yield follows an approximately linear but non-proportional behaviour for electron energies above 100 keV:

$$L \sim E_e + 11.5 \text{ keV} \qquad (2.12)$$

This relation can be used to explain the displacement of the multiple sum peaks in NaI(Tl) spectrometers. The absorption of a single gamma ray in the detector will require approximately three interactions, 2–3 Comptons and 1 photoeffect. Similarly the absorption of two coincident gamma rays will require of the order of 6 interactions and so on. If each interaction produces a shift of approximately 10 keV, then calibrating with single peaks, certainly cannot reproduce the energy of the sum peaks because of the additional interactions and consequently shifts that the second, third, etc., summed gamma-rays will generate. This will appear as shifts of approximately multiples of 30 keV depending on the multiplicity of the cascade.

References

1. Guadilla, V.: Ph.D. Thesis, University of Valencia (2017)
2. Duke, C.L., et al.: Nucl. Phys. A **151**, 609 (1970)
3. Johansen, K.H., Bonde Nielsen, K., Rudstam, G.: Nucl. Phys. A **203**, 481 (1973)
4. Erdal, B.R., Rudstam, G.: Nucl. Inst. Methods **104**, 263 (1972)
5. Bohr, A., Mottelson, B.R.: Nuclear Structure. Benjamin, Inc., New York (1969) (a) pp. 184–188; (b) pp. 211, 302
6. Bykov, A.A., Wittmann, V.D., Morozand, F.V., Naumov, Yu V.: Izv. Akad. NaukSSSR, Ser. Fiz. **44**, 918 (1980)
7. Alkhazov, et al.: Phys. Lett. B **157**, 350 (1985)
8. Hardy, J.C., et al.: Phys. Lett. **71B**, 307 (1977)
9. Firestone, R.B.: Phys. Lett. **113B**, 129 (1982)
10. Hardy, J.C., Jonson, B., Hansen, P.G.: Phys. Lett. **136B**, 331 (1984)
11. Firestone, R.B., et al.: Phys. Rev. Lett. **33**, 30 (1974)
12. Firestone, R.B., et al.: Phys. Rev. Lett. **35**, 401 (1975)
13. Firestone, R.B., et al.: Phys. Rev. C **25**, 527 (1982)
14. Greenwood, R.C., et al.: Nucl. Instr. Methods Phys. Res. A **314**, 514 (1992); Greenwood, R.C., et al.: Nucl. Instr. Methods Phys. Res. A **390**, 95 (1997)
15. Nitschke, J., et al.: GSI Ann. Rep. **1994**, 285 (1995)
16. Rubio, B., et al.: J. Phys. G Nucl. Part. Phys. **44**, 084004 (2017)
17. Tain, J.L., et al.: J. Korean Phys. Soc. **59**, 1499 (2011); Valencia, E., et al.: Phys. Rev. C. **95**, 024320 (2017)
18. Tain, J.L., et al.: Nucl. Instr. Methods Phys. Res. A **803**, 36 (2015)
19. Simon, A., et al.: Nucl. Instr. Methods Phys. Res. A **703**, 16 (2013)
20. Karny, M., et al.: Nucl. Instr. Methods Phys. Res. A **836**, 83 (2016)
21. Algora, A., et al.: Phys. Lett B **819**, 136438 (2021)
22. Agramunt, J., et al.: Nucl. Instr. Methods Phys. Res. A **807**, 69 (2016)
23. Guadilla, V., et al.: Nucl. Instr. Methods Phys. Res. A **854**, 134 (2017)
24. Cano, D., et al.: Nucl. Instr. Methods Phys. Res. A **430**, 488 (1999)
25. Guadilla, V., et al.: Nuclear Inst. Methods Phys. Res. A **910**, 79–89 (2018)
26. Jordan, D.: Ph.D. Thesis, University of Valencia (2009)
27. Estevez Aguado, M.E.: Ph.D. thesis, University of Valencia (2012)
28. Estevez Aguado, M.E.: Phys. Rev. C **84**, 034304 (2011)
29. https://www.nndc.bnl.gov/ensdf/
30. https://www-nds.iaea.org/relnsd/nubase/nubase_min.html
31. Capote, R., et al.: Nuclear Data Sheets **110**, 3107 (2009). https://www-nds.iaea.org/RIPL-3/
32. Cano, D., et al.: Nucl. Instr. Methods Phys. Res. A **430**, 333 (1999)
33. Agostinelli, S., et al.: Nucl. Instr. Methods Phys. Res. A **506**, 250 (2003)
34. Taín, J.L., Cano-Ott, D.: Nucl. Instr. Methods Phys. Res. A **571**, 728 (2007); Nucl. Instr. Methods Phys. Res. A **571**, 719 (2007)
35. Algora, A., et al.: Phys. Rev. Lett. **105**, 202501 (2010)
36. Karny, M., et al.: Nuclear Phys. A **640**, 3 (1998)
37. Rice, S., et al.: Phys. Rev. C **96**, 014320 (2017)
38. Guadilla, V., et al.: Eur. Phys. Jour. Web Conf. **146**, 10010 (2017)
39. Rasco, B.C., et al.: Phys. Rev. C **95**, 054328 (2017)
40. Rasco, B.C., et al.: JPS Conf. Proc. **6**, 030018 (2015)
41. Spyrou, A., et al.: Phys. Rev. Lett. **113**, 232502 (2014)
42. Guttormsen, M., et al.: Nucl. Instrum. Methods Phys. Res. Sect. A **255**, 518 (1987)
43. Dombos, A.C., et al.: Phys. Rev C **93**, 064317 (2016)
44. Becvar, F.: Nucl. Instr. Methods Phys. Res. A **417**, 434 (1998)
45. Engelkemeir, D.: Rev. Sci. Instr. **27**, 589 (1956)

Part II
The Total Absorption Gamma Spectroscopy Analysis in Detail

Chapter 3
Total Absorption Gamma Spectroscopy Analysis in Depth

Abstract In this chapter a more detailed discussion of the analysis techniques will be presented. First an extended description of how the response function is calculated is provided. Then the algorithms most commonly used to solve the TAGS inverse problem are presented.

3.1 Calculation of the Response

In the earlier chapter we provided a simplified description of how the response function is calculated recursively when only gamma rays are involved:

$$R_j = \sum_{k=0}^{j-1} b_{jk} g_{jk} \otimes R_k \qquad (3.1)$$

where R_j is the response to level j, g_{jk} is the response of the gamma transition from level j to level k, b_{jk} is the branching ratio for the gamma transition connecting level j to level k, and R_k is the response to level k.

This simple representation was intended to provide a picture of the essential aspects of the method. In this section we provide the full description of how the response function is calculated. Equation 3.1 should be extended to take into account all possible physical processes that are involved in the beta decay and the subsequent electromagnetic de-excitation of the level populated. We will start by changing slightly the way we represent it:

$$r_j = \sum_{k=0}^{j-1} b_{jk} g_{jk} \otimes r_k \qquad (3.2)$$

where instead of using R_j we substitute a lower case r_j for it to represent the response of the setup to the gamma cascade that deexcites level j only. A necessary improvement of Eqs. 3.1 and 3.2 requires taking into account the internal conversion process [1]. This can be incorporated easily through the replacement represented below:

© The Author(s), under exclusive license to Springer Nature Switzerland AG 2024
A. Algora et al., *Total Absorption Technique for Nuclear Structure and Applications*,
SpringerBriefs in Physics, https://doi.org/10.1007/978-3-031-58864-8_3

$$g_{jk} \rightarrow \left(\frac{1}{1+\alpha_{jk}^{tot}} g_{jk} + \frac{\alpha_{jk}^K}{1+\alpha_{jk}^{tot}} e_{jk}^K \otimes x^K + \cdots \right). \tag{3.3}$$

This is an important correction, because when internal conversion occurs, the gamma transition connecting the states should not be generated. Please note that the formula represents only the first term (X-ray emission) in the complex emission process following the conversion for the K-shell. In these cases the energy of the electromagnetic transition is given to an electron in the atomic shell[1] and leaves a vacancy. The subsequent filling of the vacancy can generate X-rays or Auger electrons, represented in the equation by the response x and e_{jk}^k. The generation of X-rays, and Auger electrons, which is represented by the second part of the Eq. 3.3 is conventionally neglected for low Z decays, since it is assumed that the low energy X-rays (and the Auger electrons) are absorbed by the dead materials around the measuring point, or by the canning of the detector. This assumption is not always valid. For example in a recently studied case [2] the large conversion coefficient of two transitions de-exciting the most strongly populated level in the decay of ^{186}Hg, and the high energies of the X-rays emitted for this high Z decay case, meant that Eq. 3.3 had to be implemented fully, in order to obtain a proper description of the measured TAGS spectrum. It should be noted that Z here always refers to the atomic number of the daughter nucleus.

The combination of (3.3) and (3.2) accounts for the response of the electromagnetic cascades that de-excite a level in the daughter nucleus, but there is still a part of the response missing. The nature of the beta decay process that takes us to the level of interest has also to be taken into account. Accordingly depending on the beta decay process β^-, β^+ or EC:

$$R_j^{\beta^-} = b^- \otimes r_j \tag{3.4}$$

$$R_j^{\beta^+} = b^+ \otimes r_j \tag{3.5}$$

$$R_j^{EC} = x^K \otimes r_j \tag{3.6}$$

Please note that b^- and b^+ represents the response of the setup to the continuous beta spectrum related to the decay to level j with an end-point energy of $Q_\beta - E_j$, where Q_β is the Q value of the beta decay and E_j is the excitation energy of the level j in the daughter nucleus.[2] The responses to b^- and b^+ differ, not only because of the annihilation of the β^+ particles with the related emission of two 511 keV gamma-rays but also due to the slightly different continuum energy distributions of the β^+ and β^- particles arising from the interaction of the emitted particles with the residual nuclear Coulomb field. The atomic relaxation after electron capture is analogous to that after internal conversion. Equation 3.4 is an approximation considering that only the K X-ray is able to interact with the detector.

---

[1] The electron is emitted with an energy $E_e = E_\gamma - E_B(E_e)$, where E_B is the binding energy of the atomic shell.

[2] In the case of β^+ decay the maximum available beta particles energy is $Q_{EC} - 2m_e c^2$.

A last comment refers to the normalisation of the response function. If coincidences with ancillary detectors are used to generate the TAGS spectrum, the normalisation of the response function for each level should agree with the total efficiency of the ancillary detector for the beta transition populating the level. This is valid for β^- and β^+ cases when coincidences with beta particles are used. A particular case is the β^+/EC case when the analysis is performed in singles, since the β^+/EC ratio has to be taken into account in the combination of the β^+ and EC responses.

3.2 Algorithms

As mentioned in the earlier chapter, the beta feeding to levels is determined by solving the equation:

$$d_i = \sum_{j=0}^{j_{max}} R_{ij}(B) f_j + C_i \tag{3.7}$$

where d_i is the content of bin i in the measured TAGS spectrum, R_{ij} is the response matrix of the TAGS setup and represents the probability that a decay that feeds level j in the level scheme of the daughter nucleus gives a count in bin i of the TAGS spectrum, f_j is our goal, namely the β feeding to the level j, and C_i represents the contribution of all contaminants to the channel i of the spectrum. The index j in the sum runs over the levels populated in the daughter nucleus in the β decay.

We also mentioned earlier that the problem represented by (3.7) is a difficult task to solve. It belongs to the category of ill-posed problems. The term ill-posed problem means here that the solution is not unequivocally determined by the data, or in other words we can have more than one solution to the problem that satisfies Eq. 3.7. For that reason it is worth exploring different options to solve it and to see the effects of applying different algorithms in the solutions obtained. This can help to identify possible systematic uncertainties in the solution related to the way the problem is solved.

Three algorithms were explored in [3]. In the following sections we will explain how the two most commonly used algorithms are implemented. These algorithms can provide only positive solutions for the feeding distribution f_i by definition. The third algorithm, the linear regularisation method, is not presented because it is less commonly used. The linear regularisation can provide negative solutions for f_i. The interested reader can see a detailed discussion of this method in [3].

3.3 Expectation Maximisation Algorithm (EM)

The expectation maximisation algorithm is a method for maximum likelihood estimation of parameters from incomplete data. The method is named after the two steps required for its application. Firstly one computes the conditional expectation of the logarithm of the likelihood. Secondly, one maximises the expectation. The method is iterative. As discussed in [3] the same algorithm is obtained from different premises using the Bayes theorem. The Bayes theorem links causes, the feedings f_j, to the effects d_i, the contents of the bins in the measured spectrum as follows:

$$P(f_j|d_i) = \frac{P(d_i|f_j)P(f_j)}{\sum_{j=1}^{m} P(d_i|f_j)P(f_j)} \tag{3.8}$$

In the Eq. 3.8 $P(f_j)$ represents the a priori probability of the feeding f_j, defined as $f_j / \sum_j f_j$, $P(d_i|f_j)$ represents the conditional probability that the datum d_i is due to the feeding f_j, which in this case is equivalent to the response function R_{ij} and $P(f_j|d_i)$ is the a posteriori conditional probability that the feeding f_j was caused by the datum d_i. As discussed in [3] the denominator $\sum_{j=1}^{m} P(d_i|f_j)P(f_j)$ guarantees the normalisation and presumes that the feedings f_j are independent.

Based on the Eq. 3.8 we can relate the expected values of the feedings $\hat{f}$ to the expected values of the data $\hat{d}$:

$$\hat{f}_j = \frac{1}{\sum_{i=1}^{n} R_{ij}} \sum_{i=1}^{n} P(f_j|d_i)\hat{d}_i, \quad j = 1,\ldots,m \tag{3.9}$$

where $\sum_{i=1}^{n} R_{ij}$ represents the efficiency to detect a decay to level j. If in Eq. 3.9 we substitute the data d_i for the expected value $\hat{d}_i$ and we insert Eq. 3.8 in 3.9 we arrive at the following equation that is solved iteratively:

$$f_j^{(s+1)} = \frac{1}{\sum_{i=1}^{n} R_{ij}} \sum_{i=1}^{n} \frac{R_{ij} f_j^{(s)} d_i}{\sum_{k=1}^{m} R_{ik} f_k^{(s)}}, \quad j = 1,\ldots,m \tag{3.10}$$

This algorithm is one of those most commonly used in our work. It has the advantage that the solutions are positive by definition, since they have a probability character. Equation 3.10 can be written in matrix form as follows:

$$f^{(s+1)} = M^s d \tag{3.11}$$

Because of the linearity of (3.11), it is possible to calculate the uncertainties and the correlations by applying the conventional techniques of error propagation at the point where the solution is reached.

$$V_f = M V_d M^T \tag{3.12}$$

where V_f and V_d represent the correlation matrices of the solution and the data respectively.

3.4   The Maximum Entropy Algorithm

Another commonly used algorithm in our analysis is the Maximum Entropy (ME). It is based on the maximum entropy principle of statistical inference [4]. The method states that the optimal solution for an inverse problem, in the sense of being less biased by our ignorance, corresponds to the one that maximises the Shannon information entropy $S[f]$ [5] that is consistent with other constraints. If the constraint applied is to reproduce the data in the least squares sense [3] the problem can be presented as follows:

$$max : S[f] - \frac{1}{\lambda}\chi^2[f] \qquad (3.13)$$

with λ as a regularisation parameter.

In this approach, the entropy represents our a priori knowledge of the solution. An equivalent way of representing (3.13) is the following:

$$min : \chi^2[f] - \lambda S[f] \qquad (3.14)$$

There are several choices for the form of the entropy functional. In [3] the following form, which has the advantage of providing positive solutions if the initial solutions are positive, was tested and implemented:

$$S[f] = -\sum_{i=1}^{n} \left(f_i ln \frac{f_i}{h_i} - f_i + h_i \right) \qquad (3.15)$$

where h_i represents an initial reference value.

The minimum condition presented in (3.14) is equivalent to solving the equation:

$$\nabla_f(\chi^2[f] - \lambda S[f]) = 0 \qquad (3.16)$$

This problem, because of the non-linear character of the entropy functional is solved numerically. In [3] an iterative algorithm similar to [6] was used after the result of the minimisation procedure:

$$\sum_{i=1}^{n} \frac{2}{\sigma_{d_i}^2} R_{ij} \left(d_i - \sum_{k=1}^{m} R_{ik} f_k \right) = \lambda ln \frac{f_j}{h_j}, \quad j = 1, \ldots, m \qquad (3.17)$$

The form of equation 3.17 suggests the use of an iterative method for the solution if one considers f_j on the left hand-side of the equation and h_j on the right hand-side as the feeding from an earlier iteration as follows:

$$f_j^{(s+1)} = f_j^{(s)} exp \left(\frac{2}{\lambda} \sum_{i=1}^{n} R_{ij} \left(d_i - \sum_{k=1}^{m} R_{ik} f_k^{(s)} \right) / \sigma_{d_i}^2 \right) \quad j = 1, \ldots, m \quad (3.18)$$

This last equation, because the result is the product of $f_j^{(s)}$ and an exponential reflects the positiveness of the next iteration solution $f_j^{(s+1)}$ if the earlier $f_j^{(s)}$ solution is positive.

A disadvantage of this method is that the solution depends on the regularisation parameter λ, and one needs to define a rule for selecting it (several are possible). As in the previous case the solution depends on the number of iterations chosen, the criteria for stopping.

According to [3] covariances can be approximated by:

$$\sigma_{f_i, f_j} = \frac{2}{\lambda^2} f_i f_j \sum_{k=1}^{n} R_{ki} R_{kj} / \sigma_{d_i}^2 \quad i, j = 1, \ldots, m \qquad (3.19)$$

This result is obtained after applying the error propagation at the point where convergence is obtained and keeping the first order term in the expansion.

3.5 Estimation of Systematic Errors

Here we provide a very brief discussion of systematic errors. As mentioned earlier in this chapter, there are ways to estimate the statistical error associated with the application of a particular algorithm to the analysis based on a given level scheme (B). This error is conventionally smaller than the error that arises from the envelope of possible solutions of the problem based on different assumed level schemes that reproduce well the measured data. This, and the impact of the errors of the constants used to normalise the different distortions of the spectrum (C_i), are taken into account in the final estimation of the errors of the analysis. Different algorithms, as the ones mentioned in this chapter can be used to solve the problem, and the differences of the solutions can help to assess on the systematic error associated to the use of a particular algorithm.

As mentioned in the earlier chapter, a reliable analysis should be preceded by a validation of the Monte Carlo code used for calculation of the response matrix $R_{ij}(B)$. This can be achieved by testing the quality of the Monte Carlo description of the setup using the spectra of known radioactive sources measured during the experiment. Once the experimental spectra of interest are well reproduced after the analysis, one can further test the quality of the branching ratio matrix B in the case of segmented spectrometers. This can be done by checking how well the experimental spectra with different multiplicity conditions or fold are reproduced with a Monte Carlo simulation that employs the results from the analysis. One might also expect that a robust analysis result should not depend on the algorithm used for the analysis. Results from different analysis algorithms can help to estimate part of the systematic errors.

References

1. Cano, D., et al.: Nucl. Instr. Methods Phys. Res. A **430**, 333 (1999)
2. Algora, A., et al.: Phys. Lett B **819**, 136438 (2021)
3. Tain, J.L., Cano, D.: Nucl. Instr. Methods Phys. Res. A **571**, 728 (2007)
4. Jaynes, E.T.: Phys. Rev. **106**, 620 (1957)
5. Shannon, C.E.: Bell Syst. Tech. J. **27**, 379 (1948)
6. Collins, D.M.: Nature **298**, 49 (1982)

Part III
Applications and Future

Chapter 4
Applications

Abstract In this chapter the most important applications of total absorption gamma spectroscopy will be presented. The first part is devoted to the impact of the technique in nuclear structure. It will include the possibility of determining the shape of the beta decaying state in particular cases and the quenching of the Gamow–Teller strength. Later in the chapter reactor applications are covered. This will include in particular the relevance of beta decay data free from the *Pandemonium* effect in reactor decay heat calculations and in the prediction of the antineutrino spectrum from reactors. Finally, examples of the use of total absorption measurements in astrophysics will be presented.

4.1 Nuclear Structure

Nuclear structure is at the heart of the applications of total absorption gamma spectroscopy. As mentioned in the second chapter, the first total absorption spectrometer was constructed and used at ISOLDE (CERN) with the aim of measuring the beta decay strengths of exotic nuclei [1]. Since those early days the technique has been applied in many regions of the nuclear chart providing relevant and interesting results.

In this framework, when we talk about nuclear structure applications, basically we think about testing the predictions of a nuclear model by comparing them with the results of a measurement. Validating model calculations is not only of relevance for improving our basic knowledge of nuclear physics, it can also be important for other applications that rely on model predictions, as will be seen later in this chapter.

For the comparison with model calculations we use the beta strength function introduced in the second chapter:

$$S_\beta(E) \equiv \frac{I_\beta(E)}{f(Z_d, Q - E)T_{1/2}} = \frac{1}{f(Z_d, Q - E)t_{1/2}} \tag{4.1}$$

where $I_\beta(E)$ is the absolute beta decay transition probability, where the total beta feeding is normalised to 1, to levels in the daughter nucleus at excitation energy E, $Q - E$ is the endpoint of the beta transition, f is the statistical rate function, and

$T_{1/2}$ is the half-life of the decay. It was shown in Chap. 1 that $S_\beta(E)$ is in practical terms the reciprocal of the ft values with $t_{1/2} = T_{1/2}/I_\beta(E)$ (partial half-life). E can represent a level in the daughter nucleus, or a group of levels included in an energy bin centred at excitation energy E.

The total absorption technique is important for all applications when the feeding distribution in the entire Q_β window has to be determined free from the *Pandemonium* effect [2]. The other two quantities appearing in the formula, namely $Q_\beta - E$ and the $T_{1/2}$ are typically obtained from independent measurements dedicated to the purpose. We will focus mainly on allowed Gamow Teller transitions, since allowed Fermi transitions are normally concentrated in a single state, and in general not strongly affected by the *Pandemonium* effect [1].

The experimental β strength is related to the experimental $B(GT)$ in the case of Gamow Teller transitions according to the following equation:

$$S_\beta(E) = \frac{1}{6147} \left(\frac{g_A}{g_V} \right)^2 \sum_E B(GT)^{exp}_{i \to f} \tag{4.2}$$

where g_A and g_V are the axial-vector and vector coupling constants. The $B(GT)$ as defined above, between the parent state and the states populated in the daughter nucleus can be related to the theoretical transition probability defined as follows:

$$B(GT)^{theo}_{i \to f} = |\langle \Psi_f | \sum_\mu \sum_k \sigma^\mu_k \, t^\pm_k | \Psi_i \rangle|^2 \tag{4.3}$$

where σ and τ are the spin and isospin operators. So now, it is possible to compare theory and experiment. The experimental $B(GT)$ deduced according to (4.2) is compared with (4.3) and this will reveal how good the model predictions are.

We will start the discussion with one application that was first introduced at ISOLDE (CERN). This was the determination of the shape of the ground state of a nucleus from a measurement of the beta decay strength in transitions to excited states in the daughter nucleus.

4.1.1 Nuclear Shape Determination

The shape is an important characteristic of the nucleus. In essence this concept is related to the equilibrium shape taken by the nucleons in the intrinsic frame. Conventionally shapes can be spherical, prolate, or oblate which are less common, etc.

---

[1] In an allowed Fermi transition, only the third component of the isospin can change and the state populated in the daughter nucleus is the Isobaric Analogue State (IAS) of the parent state. In allowed Gamow Teller transitions the isospin as well as the spin can change by one unit, and consequently a number of states can be populated in the daughter nucleus. If the Q_β energy window is large enough, the strength can be very fragmented and its measurement can be affected by *Pandemounim*

The departure from the spherical shape in a particular system has important consequences in nuclear structure. States characterised by larger deformation, have more
collective behaviour and consequently the associated electromagnetic transitions are
typically stronger and faster. Because of this relationship, and based on the rigid rotor
model, the deformation parameter β, or the related ε, is conventionally determined
by measuring the electromagnetic transition strength $B(E2) \uparrow$:

$$\beta = (4\pi/3ZR_o^2)\sqrt{(B(E2) \uparrow /e^2)} \tag{4.4}$$

here R_o represents the nuclear radius, and Z the number of protons in the nucleus.
The reader should take note that the vertical arrow indicates that the transition is
from a lower to higher excitation state.

This relationship is well defined for states belonging to a rotational band in a
nucleus. So, from the measured $B(E2)$ values the deformation of excited states can
be determined. A compilation of such values can be found in [3].

An important question that remains is how to determine the deformation of the
ground state of a nucleus. As mentioned before we can assume that the ground state is
part of the rotational band, which is characterised by the same intrinsic deformation
and apply the relation (4.4). This is not the only possibility however. Actually, the
most natural option for determining the deformation of any state including the ground
state case is the measurement of the spectroscopic quadrupole moment (Q_s) [4]
whenever that is possible. The spectroscopic quadrupole moment is defined as:

$$Q_s(I) = \langle I, m = I | Q_2^0 | I, m = I \rangle = \sqrt{\frac{I(2I-1)}{(2I+1)(2I+3)(I+1)}}(I||Q||I) \tag{4.5}$$

Here Q_2^0 is the operator of the quadrupole moment, I is the spin of the state and
m its projection. The Q_2^0 operator is defined as follows:

$$Q_2^0 = Q_z = \sum_{i=1}^{A} Q_z(i) = \sum_{i=1}^{A} e_i(3z_i^2 - r_i^2) = \sqrt{\frac{16\pi}{5}} \sum_{i=1}^{A} Ae_i r_i^2 Y_2^0(\theta_i, \phi_i) \tag{4.6}$$

which characterises the departure from spherical shape. In this last equation e_i is the
charge of a nucleon and (x_i, y_i, z_i) its Cartesian coordinates. In the last part of the
formula the operator is expressed as the zero order component of a rank 2 tensor in
polar coordinates.

Unfortunately, because of the definition of the quadrupole moment operator
(Eq. 4.5), the expectation value of the spectroscopic quadrupole moment (Q_s) for
$I = 0, 1/2$ is zero. So we can not extract the deformation information for the ground
states of even-even nuclei, characterised by spin $I = 0$, from this kind of measurement. The nucleus can have an intrinsic deformation, but we can not determine it
using (4.5).

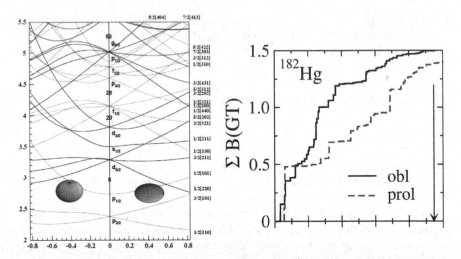

Fig. 4.1 Left panel: Nilsson model orbitals for neutrons and protons for N, Z < 50. The different ordering of the single particle orbitals depends on the sign of the shape of the nucleus. Right panel: Gamow–Teller strength calculated for the beta decay of ^{182}Hg, which shows different patterns of the beta strength depending on the assumed shape, oblate or prolate, of the ground state [9]. The arrow represents the Q value of the decay

Beta decay offers an additional possibility. This was proposed from the theoretical viewpoint by Hamamoto and collaborators [5] and further explored by Sarriguren, Petrovichi and collaborators [6, 7]. The basic idea is that the beta strength measured in the daughter nucleus is determined by the nuclear deformation of the decaying parent state. This can be understood intuitively using the Nilsson model [8]. This model provides the ordering of single particle states in a deformed nucleus. If, in the framework of the deformed shell model, we fill all the deformed orbitals corresponding to a particular deformation with nucleons, it is clear that this configuration will determine the single particle states to which those nucleons can beta decay, since the ordering of the states depends on whether we are on the prolate or oblate side of the Nilsson diagram. As a consequence, different patterns of the beta strength can occur depending on the deformation of the parent nucleus. Here the Nilsson model was just used to give an insight into the phenomenon, but in reality more complex models are required for the calculations, typically the quasi random phase approximation or QRPA model or the shell model (see right panel of Fig. 4.1).

This possibility was first explored experimentally at ISOLDE using the total absorption spectrometer *Lucrecia*. Measurements of the beta strength of nuclei using total absorption spectroscopy in the region $A \sim 80$ and their comparison with theory allowed us to determine the deformations of ^{76}Sr (see Fig. 4.2) and ^{74}Kr. The results show that ^{76}Sr is one of the most deformed $N = Z$ nuclei in its ground state [10], and that ^{74}Kr is probably a mixture of shapes in its ground state [11]. In the latter case the mixed character was deduced from the poor reproduction of the beta strength by the model. The experimental result lies between the values the calculations give

Fig. 4.2 Comparison of the experimentally determined strength with theoretical calculations for the beta decay of ^{76}Sr. The experimental values are shown in black with the grey area indicating the uncertainty. The theory shows the values for oblate(blue) and prolate(red) shapes. From the figure it can be inferred that ^{76}Sr is a deformed prolate nucleus in its ground state [10]. Copyright of Physical Review Letters (APS)

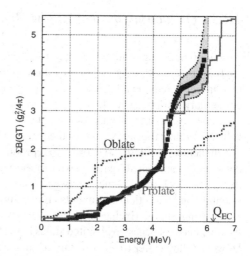

for prolate and oblate shapes, which seems to be consistent with the deduced mixed character of this ground state based on a measurement of an $E0$ transition.[2] Since the first measurements, this method has been applied to other decays in the $A \sim 80$ region [13, 14] and to nuclei in the $A \sim 190$ Pb region. In [15] the spherical character of the ground states of ^{190,192}Pb was also deduced from similar measurements. A summary of all cases measured in the $A \sim 80$ and $A \sim 190$ regions at ISOLDE is presented in the review article [16].

Another interesting case, also recently studied at ISOLDE, is the beta decay of ^{186}Hg. This case is unusual. From around mass $A = 186$ down to $A = 180$ a staggering in the nuclear radii appears in the Hg isotopes, which until very recently was unique in the nuclide chart. This staggering has been interpreted as being due to drastic shape changes starting around ^{186}Hg that depend on the mass of the isotopes. In this framework ^{186}Hg has been thought to be oblate in its ground state. But the ground state of the daughter nucleus (^{186}Au) is assumed to be prolate. So here we face an interesting case where the beta decay of ^{186}Hg connects states of different deformation and the beta strength to various excited states can be determined by small amounts of mixing of the ground state wave function. This was explored in [17], where the best description of the half-life and the pattern of the beta strength is obtained assuming mixed states with dominant prolate character for both ^{186}Hg and ^{186}Au in contradiction to the assumed oblate deformation of ^{186}Hg. This problem should be further explored both experimentally and theoretically.

The method of inferring the shape from the beta strength in the daughter, has also been applied in recent publications by the SUN collaboration [18].

---

[2] Electromagnetic transitions of monopole character (E0) are not expected between states with very different intrinsic shapes. For that reason the existence of $E0$ transitions between co-existing states is considered to be an indication of the mixing of those states [12].

4.1.2   Quenching of the Gamow–Teller Strength

Another interesting topic of research related to total absorption measurements is the
study of the so-called quenching of the Gamow–Teller strength. This is a problem
that has attracted considerable attention for more than four decades and it is not yet
completely settled [19–22].

It turns out that the beta strength obtained experimentally seems to be systematically lower than the predictions of theory. This has been referred to in the literature as
the quenching problem. So in order to reproduce the experimental data the $g_A = 1.27$
constant is "quenched" with respect to the value used to describe the free neutron
decay. The lack of agreement between theory and experiment demands both experimental and theoretical arguments. Experimentally, one can address this problem
either using Charge Exchange reactions or β decay. Charge Exchange reactions have
the advantage of accessing excited states in the final nucleus at high excitation energy.
However, they suffer from uncertainties in the reaction mechanisms as well as in the
interpretation of the background. These two difficulties do not affect the β decay. In
β decay the main difficulties are the limited energy window available in the process,
and the possible *Pandemonium* effect. Considering this second problem, the comparison is only meaningful if experimental data from total absorption experiments,
free of *Pandemonium*, are employed.

Regarding the limited energy available in the decay, there are two regions in
the nuclide chart of particular interest: the nuclei around ^{100}Sn and the rare-earth
nuclei above ^{146}Gd. The reason is that in both cases, the Gamow–Teller strength
is dominated by one main component, namely the $\pi g_{9/2} \rightarrow \nu g_{7/2}$ transition in the
^{100}Sn region and the $\pi h_{11/2} \rightarrow \nu h_{9/2}$ transition in the rare-earth case. All other proton
occupied orbitals have no corresponding empty neutron orbital partner (see Fig. 4.3
for an schematic picture in the ^{100}Sn region). The condition, that the Gamow–Teller
decay should be dominated by almost pure transitions at relatively low energy in the
daughter, makes these two regions ideal for such studies.

A relatively simple theoretical calculation of the expected β strength on the β^+
side was carried out by Towner [23] for decays in both regions of the Segrè Chart.
In Towner's work a hindrance factor h is defined as the ratio of the summed GT
strengths in theory and experiment. As a first approach he adopted the so-called
extreme single particle approach $(s.p)$, where one considers only the two pairs of
orbitals, $\pi g_{9/2}$-$\nu g_{7/2}$ and $\pi h_{11/2}$-$\nu h_{9/2}$ as pure configurations. Then a series of successive corrections to this approach were made taking into account additional effects
like pairing, core polarisation and higher-order effects. Towner then compared how
hindered the corrected theoretical strengths would be in relation to the extreme $s.p$
picture. This result defines a theoretical hindrance factor (or ratio) that can be compared later with the hindrance obtained from the ratio of the extreme single particle
approximation and experiment. The theoretical hindrance was evaluated for the range
of $n = 1$ to $n = 10$ active protons in the $g_{9/2}$ orbital in the ^{100}Sn region and $n = 1$ to
$n = 12$ in the $h_{11/2}$ orbital in the ^{146}Gd region. Both regions have been studied with
the total absorption spectrometer [25], originally developed at the Berkeley Super-

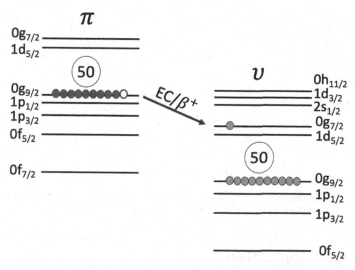

Fig. 4.3 Schematic picture of the relevant orbitals in the beta decay of ^{100}Sn

HILAC, at the GSI Mass SEParator (MSEP) [24]. The results of these measurements can be seen in Refs. [26] (^{100}In), [27] (^{97}Ag), [28] (^{148}Dy), [29, 30] (^{150}Ho) and [31] (^{148}Tb(2^- and 9^+ isomers), and ^{152}Tm(2^- and 9^+ isomers)). Summaries of the results and comparisons of the hindrances can be found in [22, 31]. In Fig. 4.4 a comparison is presented for the Gd region cases.

In this context the beta decay of ^{100}Sn is special, since the expectation in this case is that all of the GT strength is concentrated in the transition to a single 1^+ state in the daughter nucleus. This decay represents the fastest allowed Gamow–Teller transition in the nuclear chart [5] and this is the reason why it has always attracted considerable attention. The relevance of this decay in the framework of the study of the quenching of the Gamow–Teller strength has been discussed recently by Gysbers et al. [21]. This decay played a key role in the comparison of their theoretical calculations, that aim for a deeper understanding of the origin of this long-standing problem, with experiment [21]. For the ^{100}Sn case it was only possible to compare their calculations with measurements made with Ge detectors, see Hinke et al. [34] and the more recent work by Lubos et al. [35]. From the total absorption perspective only the extrapolations from the work of Batist et al. [32] are available for ^{100}Sn until a TAGS study of this special decay is completed. It is worth mentioning that according to Gysbers et al. [21], the quenching arises to a large extent from the coupling of the weak force to two nucleons as well as from strong correlations in the nucleus. A nice agreement was obtained in [21] for some of the models studied with the extrapolations of Batist et al. [32] but work is still needed here.

In this context clearly the total absorption technique is also of great importance from the experimental point of view. A new experiment was partially performed at RIKEN [33] aiming at studying of the beta decay of ^{100}Sn. The proposal presently awaits completion.

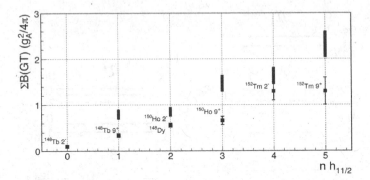

Fig. 4.4 B(GT) sum obtained from the TAGS measurements presented in [31] in comparison with the theoretical calculations by [23]. Copyright of Physical Review C (APS)

4.2 Reactor Applications

In nuclear reactors we use fission to produce energy. Beta decay is also of great relevance in this framework: every fission is followed by approximately six beta decays. This can be easily understood, since the fission products are produced with approximately the proton to neutron ratio corresponding to the fissioning system $A + 1n$, in the case of neutron induced fission of a nucleus with mass number A. Since this ratio differs from the proton to neutron ratio at the line of nuclear stability for the given fragment mass, in order to reach stability several beta decays must occur. This is presented schematically in Fig. 4.5.

The relatively large number of beta decays that occur after each fission has several consequences. One is that in a working reactor a non-negligible amount of the released energy is related to the beta decay of the fission products. Please note that in this estimation the energy of the antineutrinos that leave the reactor vessel is not considered. When the reactor is shut down, the beta decay of the fission products becomes the most relevant source of energy. So accounting for the beta decay of the fission products is a relevant issue for the security of nuclear reactors and for the safe storage of spent fuel. Table 4.1 shows the mean energy released in the fission of the most common nuclear fuels ^{235}U and ^{239}Pu. The second important consequence is that reactors are the strongest man-made pacific sources of antineutrinos, since each beta decay is accompanied by the emission of one antineutrino. In the following subsections we will discuss the relevance of the study of beta decays from the perspective of these two aspects.

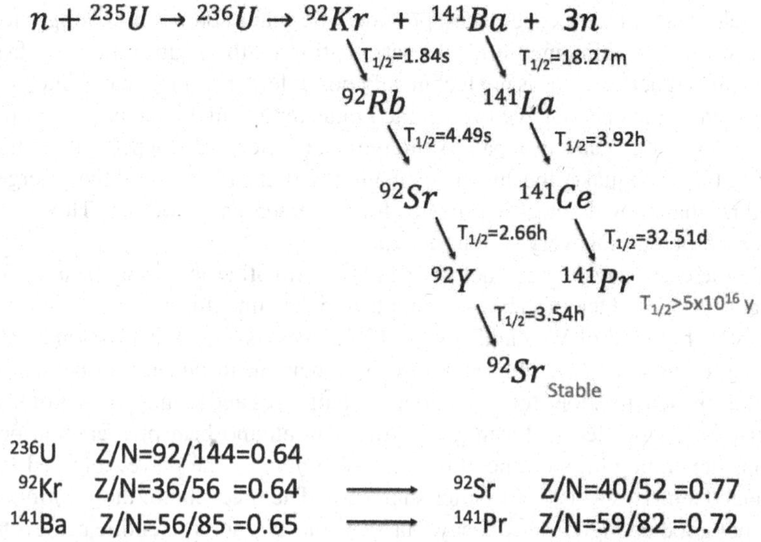

$$n + {}^{235}U \rightarrow {}^{236}U \rightarrow {}^{92}Kr + {}^{141}Ba + 3n$$

Fig. 4.5 The neutron excess of fission products and its relation to the number of decays after each fission for a common fission channel. In the lower part of the figure the neutron to proton ratio of ^{236}U and the direct fission products are presented. The neutron to proton ratios of the corresponding stable nuclei at the end of the decay chains are also shown

Table 4.1 Energy released in and after fission of the most important fissile isotopes ^{235}U and ^{239}Pu. The values are given in MeV/fission [36]

Contribution	^{235}U	^{239}Pu
Fission products kinetic energy	166.2(13)	172.8(19)
Prompt neutrons	4.8(1)	5.9(1)
Prompt gamma rays	8.0(8)	7.7(14)
Beta energy of fission fragments	7.0(4)	6.1(6)
Gamma energy of fission fragments	7.2(13)	6.1(13)
Subtotal	192.9(5)	198.5(8)
Energy taken by the neutrinos	9.6(5)	8.6(7)
Total	202.7(1)	207.2(3)

4.2.1 Decay Heat

Decay heat is defined as the amount of energy released by the decay of radioactive nuclei produced in the reactor. A precise estimate of this energy release is important for the safety assessment of any type of nuclear plant, the handling of the fuel, the design and transport of fuel-storage flasks, and for radioactive waste management. The decay heat amounts to about 7% of the released energy in a typical working nuclear reactor, but it depends on the fuel used. Once the reactor is shutdown, the energy released in radioactive decay provides the main source of heating. Hence, cooling has to be maintained after the termination of the neutron-induced fission processes in a reactor. If this is not technically possible, serious nuclear accidents can happen such as the one that occurred at the Fukushima Daiichi nuclear power plant, following a great tsunami in Japan. Although the reactor had stopped automatically following the earthquake, the ensuing tsunami flooded and damaged the emergency generators, that provide electrical power for the emergency cooling. This was the main reason behind this very serious accident.

Since the early days of the nuclear industry, estimating the decay heat has been a relevant question. One of the first attempts to determine this relevant quantity can be found in the work of Way and Wigner [37]. They assumed that fission products constitute a statistical assembly and based on their mean nuclear properties, they deduced empirical relations for the radioactive half-lives and atomic masses of fission products. This work led to formulae for the gamma and beta plus gamma power functions depending on the time after shutdown, namely the power released in the time range from 10 s to 100 days after shutdown. The precision of these formulae is clearly not good enough for use today, but Way and Wigner's scientific achievement is remarkable, considering the paucity of nuclear data at the time. The interested reader is encouraged to read this seminal work.

Nowadays the most common way of calculating the decay heat is based on summation calculations:

$$f(t) = \sum_i (\overline{E}_{\beta,i} + \overline{E}_{\gamma,i} + \overline{E}_{\alpha,i}) \lambda_i N_i(t) \tag{4.7}$$

where $\overline{E}_i$ is the mean decay energy of the ith nuclide (β or charged-particle, γ or electromagnetic and α or heavy particle components), λ_i is the decay constant of the ith nuclide, and $N_i(t)$ is the number of nuclides of type i at the cooling time t. Here E_α has been added for completeness, but the contribution of this component is small in a working reactor and at short cooling times after shutdown. Formula (4.7) means summing the decay activities ($\lambda_i N_i$) times the mean energies $\overline{E}_i$ released per decay to obtain the power function $f(t)$. This is the reason why the method is called a summation calculation. The gamma and beta energy separation is useful for radiological reasons, since the gamma radiation is more penetrating in general than the beta part.

For calculations using (4.7) we need a large amount of nuclear data, since in fission several hundred fission products are created. λ_i can be obtained from the half-life of the decay using the relation: $\lambda = ln(2)/T_{1/2}$. The number of nuclides at time t ($N_i(t)$) requires that we solve a linear system of coupled first order differential equations, (see the appendix) and the mean energies released per decay $\overline{E}$ are also needed.

If the level schemes populated in the beta decays are known, the mean energies released per decay can be estimated as follows:

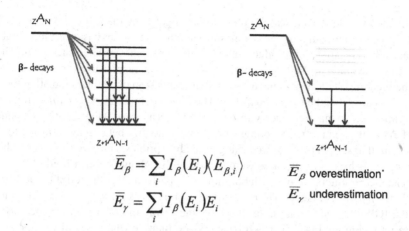

$$\overline{E}_\beta = \sum_i I_\beta(E_i)\langle E_{\beta,i}\rangle \qquad \overline{E}_\beta \text{ overestimation}^\cdot$$

$$\overline{E}_\gamma = \sum_i I_\beta(E_i)E_i \qquad \overline{E}_\gamma \text{ underestimation}$$

Fig. 4.6 Impact of the *Pandemonium* effect in the determination of the mean energies. Left panel: real situation, right panel: case suffering from *Pandemonium*

$$\overline{E}_\gamma = \sum_j I_{\beta j} E_j \qquad (4.8)$$

$$\overline{E}_\beta = \sum_j I_{\beta j} < E_\beta >_j \qquad (4.9)$$

where $I_{\beta j}$ represents the beta decay probability to level j, where the beta feeding is normalised to 1, and E_j is the excitation energy of the level in the daughter nucleus. The $\overline{E}_\gamma$ estimation is based on the assumption that each level reached by beta decay de-excites by gamma transitions. It should be noted that in more elaborate calculations the conversion electron process is also taken into account. The $\overline{E}_\beta$ determination is more complex. It requires the calculation of the mean $< E_\beta >_j$ beta energy corresponding to each endpoint $Q - E_j$, and requires the making of assumptions about the shape of the beta transitions.

At this stage it should be obvious that large amounts of beta decay data are required for the calculations.[3] The necessary data are compiled and evaluated by international collaborations or national projects such as JEFF, JENDL, ENDF, etc. [38–41].

Equations 4.8 and 4.9 require that the feeding distribution in the individual beta decays is properly determined, and this is why total absorption measurements are highly relevant for the application of the summation method. Clearly if we use a beta decay level scheme that suffers from the *Pandemonium* effect, the gamma mean energy deduced will be underestimated and the mean beta energy will be overestimated (see Fig. 4.6).

---

[3] Additional nuclear data are also required, such as fission yields and neutron capture cross-sections, in order to determine the inventory of nuclides N_i (see appendix).

In 2004 we started a research programme triggered by the work of Yoshida et al. [42]. They were investigating how well summation calculations based on the available databases were able to reproduce integral measurements of the decay heat. They identified several relevant fission products that could be responsible for a discrepancy in the predictions of the gamma decay heat in the fission of ^{239}Pu, in the 300–3000 s range, because their beta decay data could be affected by the *Pandemonium* effect. At that time we also made contact with staff in the International Atomic Energy Agency (IAEA), that set in motion meetings of experts to identify beta decays relevant to the problem, that should be measured using total absorption spectroscopy. As a result of the meetings high priority cases were identified that should be measured [43].

The measurements of our collaboration were performed at the IGISOL facility of the University of Jyväskylä. The reason for selecting this facility was the development of the IGISOL fission ion guide. It allows the extraction of fission products using a He jet technique [44]. This is relevant, since many of the cases of interest were isotopes of refractory elements, that are difficult to extract from conventional ion sources. In the typical experiment at IGISOL, a natural U target is bombarded with a proton beam, that induces fission. The fission products are extracted using the fission ion guide, accelerated and separated in the IGISOL mass separator and then further separated using the JYFL Penning trap [45]. The JYFL Penning trap is a device that was developed for precise mass measurements, but it can also be used as a high resolution mass separator for spectroscopic measurements. The isotopically separated beam after the Penning Trap is then transported to the measuring point. This further mass separation is important in calorimetric measurements, since we can use isotopically clean beams for the measurements and reduce the effect of contaminants and the associated subtractions in the analysis.

Figure 4.7 shows the impact of 7 nuclides measured in our second experiment performed at Jyväskylä [46–48]. The seven decays were considered to be of high-priority in the lists published in the IAEA reports [43], but only five of them (^{104,105,106,107}Tc and ^{105}Mo) were found to suffer from the *Pandemonium* effect. This clearly emphasises the need to perform the measurements. Two cases, ^{102}Tc and ^{101}Nb, that were dominated by large ground state to ground state feedings, do not suffer seriously from the *Pandemonium* effect. In Table 4.2 we present a comparison of the mean energies deduced from our total absorption measurements for these seven decays as an example.

Since then additional measurements have been performed that have contributed to an improvement in the beta decay data used in decay heat summation calculations. The results have also been important in other contexts as we shall see in the next subsection. The impact of all the published results performed by the Valencia–Nantes–Surrey–Jyväskylä collaboration has been presented in [49] (see Fig. 4.8). The situation has improved dramatically for ^{239}Pu, but there is still room for improvement for ^{235}U.

Other groups are also working on this topic [50, 51] and new measurements using further purification with ion-manipulation techniques have recently been performed at Argonne National Lab using the MTAS detector [52].

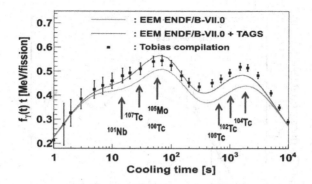

Fig. 4.7 Impact of seven decays [46–48] on the summation calculations of the gamma component of the decay heat of ^{239}Pu compared with standards in the field (Tobias compilation). The summation calculations are performed using the ENDF/B-VII.0 database, where in one case (ENDF/B-VII.0 + TAGS) the mean gamma and beta energies of the seven decays are substituted with the mean energies deduced from the TAGS measurements. The arrows show the maximum of their contribution on the time axis. Copyright of Physical Review Letters (APS)

Table 4.2 Comparison of mean gamma and beta energies included in the ENDF/B-VII database, from the year 2012, with the results obtained from our TAGS measurements (in keV) [46–48]. The errors are primarily determined by the various assumptions on the level schemes and parameters of the statistical model. The ENDF/B-VII database data were based on Ge detector measurements

Nuclide	$T_{1/2}$	$\overline{E_\gamma}$	$\overline{E_\gamma}$	$\overline{E_\beta}$	$\overline{E_\beta}$
	s	ENDF	TAGS	ENDF	TAGS
^{101}Nb	7.1(3)	270(22)	445(279)	1966(307)	1797(133)
^{105}Mo	35.6(16)	552(24)	2407(93)	1922(122)	1049(44)
^{102}Tc	5.28(15)	81(5)	106(23)	1945(16)	1935(11)
^{104}Tc	1098(18)	1890(31)	3229(24)	1595(75)	931(10)
^{105}Tc	456(6)	668(19)	1825(174)	1310(205)	764(81)
^{106}Tc	35.6(6)	2191(51)	3132(70)	1906(67)	1457(30)
^{107}Tc	21.2(2)	515(11)	1822(450)	2054(254)	1263(212)

4.2.2 Reactor Antineutrinos

Neutrino physics is a fascinating field of growing importance. Understanding these elusive particles is not only important for our basic knowledge of the building blocks of matter. It is also of importance because neutrinos can act as messengers providing information about harsh stellar environments of almost impossible access. Consider for example how to obtain information about processes that occur in the Sun or during a supernova explosion. Neutrinos are produced in such environments and detecting them can provide useful information to validate our models of what happens in the prevailing conditions there.

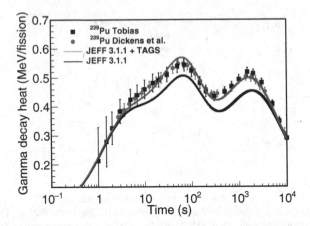

Fig. 4.8 Impact of all the TAGS measurements, performed by our collaboration and published up to 2020, on the summation calculations of the gamma component of the decay heat of ^{239}Pu compared with standards in the field [49]. Copyright of EPS A. In this comparison the JEFF 3.1.1 database was used and it did not include TAGS data for the relevant nuclei [53] and only included the earlier measurements by Greenwood et al. The decay heat calculations are courtesy of Lydie Giot [54]

There are however inherent difficulties. Detecting neutrinos is challenging, because of their very small interaction cross-section[4] with matter, typically of the order of 10^{-38} cm^2. Neutrinos are also very special particles, since it has been shown that they can oscillate. They can be produced as one type but they can then change in time into another. One of the most important goals of neutrino physics is to determine the parameters that govern neutrino oscillations. The reader will find a primer on neutrino oscillations in the appendix.

Reactors, as mentioned earlier in this chapter, are copious sources of antineutrinos, so they can be used for such studies. This idea was put forward already in the 1980s, see for example [55, 56]. A number of collaborations including those at the Daya Bay, Double Chooz and RENO reactors have recently measured the antineutrino spectrum from reactors and have determined the oscillation parameter θ_{13}. It had been assumed that this parameter was small, and it came as a surprise in 2012 that it was not as small as expected. The θ_{13} parameter was one of the parameters missing that are necessary to improve the picture of the neutrino oscillation pattern. It should be noted that other oscillation parameters are obtained from measurements of atmospheric and solar neutrinos. The result that θ_{13} is not so small, paves the way for the determination of other relevant parameters, such as the CP phase factor, using reactors. Details can be found in [57–60].

---

[4] Cross section is a measure of the reaction or interaction probability. It is measured in units of area. For example typical nuclear reactions have cross sections of the order of 10^{-24} cm^2. In nuclear and particle physics, the conventional cross section unit is the barn b, where $1\ b = 10^{-28}$ m^2 = 10^{-24} cm^2.

Understanding the primary antineutrino spectrum generated in the core of the reactors is relevant for these experiments, mainly because some of the detectors used to measure the antineutrino flux are not always precisely located in the isoflux lines of the several reactors of the power plant.

There are presently two alternative methods of calculating the antineutrino spectrum from reactors. One possibility, the conversion method, is based on the transformation of the integral measurements of the beta spectrum of fission products performed at Institut Laue Langevin (ILL) by Schreckenbach and collaborators [61]. In these measurements thin targets of ^{235}U and ^{239,241}Pu were bombarded with thermal neutrons from the reactor and the resulting aggregate beta spectra of all fission products were measured with the BILL magnetic spectrometer [62]. Since the measurements are integral measurements, it is not a trivial task to disentangle the contribution of the different individual beta branches, so for the conversion of the beta spectrum into the antineutrino spectrum several assumptions have to be made. The conversion is based on the conservation of energy since the transition energy is shared by the electron and the antineutrino, but the problem is that this can be done properly only for individual transitions and not for the aggregate spectra. Schreckenbach et al., assumed that there were 30 beta decays branches (with effective endpoints) to convert the observed beta energy spectrum into the antineutrino spectrum. This procedure was revisited independently by Mueller et al. [63] and Huber [64], who reanalysed the data from [61] incorporating improvements that were possible thanks to the large amount of nuclear data collected since the earlier work by Schreckenbach et al.[5] The upgraded Huber–Mueller model has become the standard in the field, but subsequent experiments have revealed surprises. In 2011 the comparison of the newly converted antineutrino spectrum with the measurements performed at short distances from the reactors showed that there is a deficit in the number of antineutrinos detected when compared with the calculated number. This result became known as the reactor antineutrino anomaly [65] and has attracted considerable attention since then. The interest was enhanced because one possible explanation of the deficit is the existence of a sterile neutrino, since an oscillation into this sterile flavour would exclude its possible detection[6] and explain the loss of flux (see Fig. 4.9). This explanation is not the only possible one, since there are many possible effects that can contribute to the deficit. The topic became even more interesting after the comparisons of the spectrum measured by the different collaborations with the calculations that showed that there is not only a normalisation problem (the 6% deficit), but also a shape distortion around 4–6 MeV. This distortion of the spectrum is colloquially known as "the bump" and at the time of writing it is not yet understood.

The main alternative to the Huber–Mueller model in calculating the reactor antineutrino spectrum is the summation method, which is very similar in nature to the method discussed in the earlier subsection on calculating the decay heat. The beta or antineutrino spectrum per fission of a fissile material from a working reactor

---

[5] The Mueller model is a summation-conversion hybrid model.

[6] Neutrinos are produced and detected according to their flavour. For example an electron neutrino can only be detected in a nuclear reaction or process sensitive to electron neutrinos.

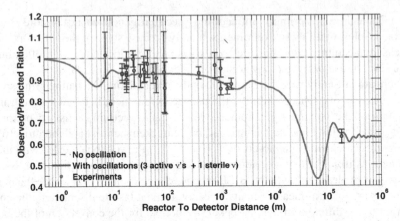

Fig. 4.9 Comparison of the calculated neutrino flux with measurements at short baselines. A possible explanation of the discrepancy of ~6% is possible assuming the existence of an sterile neutrino. Figure courtesy of Thierry Lasserre [66]

can be calculated as the sum of the appropriate spectra of all the fission products weighted by their activities:

$$S_k(E) = \sum_i \lambda_i N_i(t) S_i(E) \tag{4.10}$$

Here k stands for the index or identifier of the fissile material and i runs over all fission products. The $\beta/\bar{\nu}$ spectrum of one fission product (S_i) is the sum of all individual transition β decay spectra (or $\bar{\nu}$ spectra), (S_i^b in Eq. 4.11), from the parent nucleus to the daughter nucleus weighted by their respective β-branching ratios (β transition probabilities) according to:

$$S_i(E) = \sum_{b=1}^{N_b} I_{\beta i}^b S_i^b (Z_i, A_i, E_{0i}^b, E) \tag{4.11}$$

where $I_{\beta i}^b$ stands for the β-transition probability of the β branch, Z_i and A_i are the atomic number and the mass number of the daughter nucleus respectively, E_{0i}^b is the endpoint of the β transition b and N_b is the number of beta branches.

Again we can expect a similar decay pattern as discussed in the framework of the decay heat calculations. Decay data used in summation calculations should be free from the *Pandemonium* effect, otherwise the calculated β decay spectra (or $\bar{\nu}$ spectra) will be distorted in cases where *Pandemonium* occurs. This again shows that total absorption measurements are necessary for the application.

In [67] the impact of the decays that are of relevance in the calculation of the gamma decay heat in ^{239}Pu (see Fig. 4.7 and Table 4.2) was evaluated for the antineutrino spectrum, showing also that they can have impacts of the order of a few % for the different nuclear fuels reaching up to 8% for ^{239}Pu. An evaluation of the most relevant cases that should be measured using the total absorption technique was first performed by the Subatech group (Nantes, France) (see [68]) and then by the Brookhaven National Laboratory (BNL) team [69]. The results of those evaluations led to a high priority list of cases that should be measured using the total absorption technique. This list was published later in the framework of a TAGS consultants meeting held by the Nuclear Data Section of the International Atomic Energy Agency [70].

Among the cases identified, the beta decay of ^{92}Rb plays a prominent role. This decay contributes up to 16% of the antineutrino spectrum emitted by a pressurised water reactor (PWR) in the energy range between 5 and 8 MeV, making ^{92}Rb the most important decay to be studied to provide a solid basis for summation calculations. This case was studied in an experiment performed at the IGISOL facility in Jyväskylä. The results showed that this decay did not suffer from the *Pandemonium* effect, but the results of the total absorption measurement determined the ground state to ground state feeding for this important decay [68], which was uncertain in previous evaluations. It should be noted that the high sensitivity of the total absorption spectrometer to the penetration of the beta particles and associated bremsstrahlung in this experiment was used to determine the ground state to ground state feeding. This case was analysed by the Subatech group [68]. The large ground state feeding was later confirmed by the measurements of the Oak Ridge group using the same technique, but a different total absorption spectrometer [51].

In this context it is worth noting that in fission several hundred fission products are produced, and we do not have total absorption measurements for all the necessary decays. The Nantes group identified decay data not suffering from the *Pandemonium* effect and incorporated them into their antineutrino summation database in order of priority, giving the highest preference to decays where there had been total absorption measurements. Their antineutrino summation database included the total absorption measurements performed by Greenwood et al. from the 1990s [71], the recent total absorption measurements performed by our collaboration and the beta spectrum shape measurements of Tengblad et al. [72], since these last measurements do not rely on gamma detection and in principle should not suffer from the *Pandemonium* effect. These techniques, total absorption and the direct beta spectrum measurements, represent the available *Pandemonium* free methods. For the very short-lived nuclei theoretical calculations employing the gross theory of beta decay were employed in the summation calculations [73].

In 2019 an updated summation calculation was performed that included all total absorption measurement results published by our collaboration [74]. These results showed an impressive improvement in relation to the earlier summation calculations of [67] and provide a description not only comparable in quality to the Muller–Huber conversion model, but in better agreement with the measurements of the antineutrino spectrum published by the Daya-Bay collaboration in the range of 2–5 MeV that dominates the antineutrino flux. The results of the summation model lie within approximately 2% of the Daya Bay results (Fig. 4.10). The results from Daya Bay have been analysed more precisely (An et al.) with the conclusion that "the SM2018

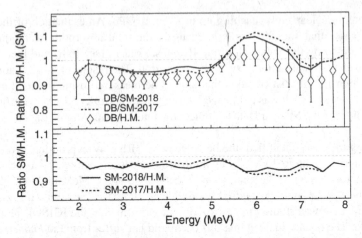

Fig. 4.10 Ratio of the Daya Bay antineutrino spectrum and the calculated spectrum using the summation model (lines). Diamonds represent the ratio of the Daya Bay and the Huber–Mueller model. The closer the ratio is to 1 the better the model description. The new summation model (SM-2018) describes slightly better the region between 2 and 5 MeV that dominates the antineutrino flux. More details may be found in [74]. Copyright of Physical Review Letters (APS)

prediction of the total IBD yield evolution is found to be more compatible with the data than the HM model". This result questions the very existence of the antineutrino reactor anomaly and shows the relevance of the total absorption technique in antineutrino summation calculations. Actually this is not the only result that questions the anomaly. The Daya-Bay collaboration was able to disentangle the contributions of the main fissile species in the observed antineutrino spectrum and showed that the discrepancy can be related to the ^{235}U data pointing to a possible normalisation problem in the ILL measurements [75] for this fissile nucleus. These results seems to be confirmed by recent measurements by Koipekin et al. [76].

Despite the considerable progress achieved in summation calculations of the antineutrino spectrum, all of the problems are not yet resolved. There are still open questions related to the shapes of the beta transitions of allowed or forbidden character and their impact [77, 78] in the summation and conversion methods. In addition the impact of the fission yields in the summation calculations are also relevant [79]. Other groups are also involved in total absorption measurements and systematic studies as can be seen for example in [50]. As a result, we expect that the summation method will be improved even further in the future. Another remaining problem is the assessment of the uncertainties associated with the method.

The summation method is important because of its versatility. It can be applied to fissioning systems for which integral measurements are not available [67] and can be helpful in providing insights into the role of individual decays [69]. The method can also be relevant in possible non-proliferation applications in which the goal is to identify illegal manipulation of the fuel in the reactor by measuring the antineutrino spectrum with modular detectors without the need to access the reactor building or naval vessel if the motive power for the ship is a nuclear reactor.

4.3 Nuclear Astrophysics

Gamma calorimeters, like the total absorption spectrometers discussed in this book, have been used in nuclear astrophysics studies for direct measurements of radiative capture reaction cross sections initiated both by neutrons or light ions. Such information is the key to understand astrophysical processes such as the s process [80], the p (or γ) process [81] and light element synthesis processes [82]. The advantage of the use of such detectors is evident, since the use of a gamma calorimeter in general is equivalent to employing a detector with a high gamma detection efficiency which can be useful when measuring processes of low probability. Examples of such detectors are the n_TOF collaboration 4π calorimeter [83], SUN [84], etc. The goal of this chapter is not to discuss such results, but to present examples of beta decay studies using the total absorption technique, that can have an impact on astrophysical studies.

Beta decay plays a key role in our understanding of fundamental astrophysical processes, like the rapid neutron capture and rapid proton capture processes (the r and rp processes). In these processes, the competition between beta decay and nuclear reactions determines how the matter flows in the nucleosynthesis network. As a result this determines the abundance of the elements. Here as an example we choose the rapid neutron capture process or r process to illustrate what happens in these processes.

Core collapse supernovaé and neutron star mergers have been considered as possible astrophysical sites for the r process. In this process a huge instantaneous neutron flux creates, by successive neutron captures, very neutron-rich nuclei that then beta decay towards stability. Approximately half of the observed abundance of elements heavier than Fe is assumed to be synthesised in this way. The recent observation of gravitational waves and the subsequent emission of electromagnetic radiation, confirmed that neutron star mergers are a site for element synthesis [85], but much still remains to be done in relation to the understanding of the possible astrophysical scenarios and their role in determining the final element abundances [86]. Crucial for this comprehension is to have proper astrophysical models of the r process, which require nuclear physics input data.

Relevant input parameters in r process model calculations are neutron capture (n,γ) reaction cross-sections and the decay properties of neutron-rich nuclei, more specifically half-lives $(T_{1/2})$ and β-delayed neutron emission probabilities (P_n) that control the matter flow in the nucleosynthesis process. The relevance of half-lives is clear, since they determine the rate of decay of the nuclei produced. As we move away from the stability line on the neutron-rich side of the nuclide chart, the neutron separation energy (S_n) gradually becomes smaller than the decay Q_β value and neutron emission from neutron unbound states populated in the decay occurs. The r process involves very neutron-rich nuclei, so beta-delayed neutron emission modifies the decay path, shifting it to lower A values and provides additional neutrons for late captures modifying the final abundance distribution [87]. This shows the importance of the P_n values. In spite of the most recent developments in radioactive beam facilities around the World that aim to allow us to reach the most exotic

nuclei and determine this information experimentally, most of the nuclei involved in
the r process calculations cannot be accessed in the laboratory today and theoretical
estimates for the $T_{1/2}$ and P_n values are required for the network calculations.

The relevance of total absorption measurements here is that both quantities are
derived from the β strength distribution $S_\beta(E)$.

$$\frac{1}{T_{1/2}} = \int_0^{Q_\beta} S_\beta(E) f(Q_\beta - E) dE \tag{4.12}$$

$$P_n = T_{1/2} \int_{S_n}^{Q_\beta} \frac{\Gamma_n}{\Gamma_n + \Gamma_\gamma} S_\beta(E) f(Q_\beta - E) dE \tag{4.13}$$

Equation 4.13 includes the competition between neutron (Γ_n) and γ emission
(Γ_γ), where Γ_γ and Γ_n represent the radiation and neutron widths. In many models it
is assumed that above Sn, neutron emission prevails and the competition is neglected
($\frac{\Gamma_n}{\Gamma_n + \Gamma_\gamma} = 1$). Recent efforts to improve the situation can be found in [88] where the
gamma and neutron emission competition above the neutron separation energy is
calculated in the framework of the Hauser-Feshbach model.

Validation of the theoretical β strength distribution $S_\beta(E)$ predicted by nuclear
models is then very relevant in this context and total absorption measurements can
provide the necessary experimental input. As an example the decay of ^{105}Mo [89],
which was studied in the framework of [46] is presented in Fig. 4.11.

As discussed in the nuclear structure section of this chapter, the assumption of a
particular shape for the parent nucleus can be important in determining the results
of the theoretical β strength distribution in the daughter nucleus. In this particular
case, the beta strength of the decay of ^{105}Mo was calculated using the FRDM-QRPA
model [46, 90]. The best theoretical description of this decay is obtained assuming a
ground state deformation of $\varepsilon_2 = -0.31$ for ^{105}Mo. The beta decay half-life is 35.6 s,

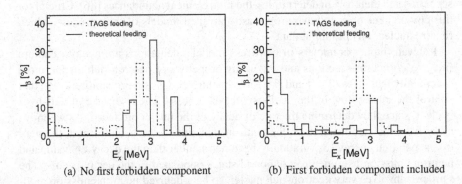

(a) No first forbidden component   (b) First forbidden component included

Fig. 4.11 Comparison of the experimentally deduced β feeding in the decay of ^{105}Mo with the
results of theoretical calculations using the FRDM-QRPA model [46, 49, 90]. The left panel was
obtained assuming only allowed GT transitions. The right panel shows the comparison with calculations that also include the first forbidden transitions. See more detail in the text

and this experimental value can be better reproduced if first forbidden transitions are included in the model calculation ($T_{1/2}^{theo} = 30.3$ s), but in that case, the experimental β intensity distribution is not reproduced so well. This can be seen in Fig. 4.11 where the experimental feeding distribution is compared with the deduced theoretical distributions. The comparisons were made with theoretical results both with and without first forbidden transitions. A better reproduction of the theoretical β feeding distribution is obtained if no first forbidden component is included in the model. However the experimental half-life is then not reproduced so well ($T_{1/2}^{theo} = 150$ s).

This shows the importance of comparing the full theoretical strength distribution or equivalently the deduced beta intensity with reliable experimental data, such as that provided by TAGS measurements, in addition to comparing the experimental half-life when assessing different models. Many models used in astrophysical applications are validated only by comparison with predicted half-lives, and this situation should be improved. If we take into account only the description of the half-life by the model, we might conclude that in the ^{105}Mo case, a first forbidden component is needed in the description of the beta decay, which in reality does not reproduce well the measured β feeding. It should be noted that the ^{105}Mo decay occurs in a region which is dominated by shape effects and where triaxiality[7] can play a role, so it might not be a simple case. As we mentioned earlier, QRPA calculations assume that both the parent and the daughter have the same deformation, which might not always be applicable in regions where shape changes among neighbouring nuclei are common. However independently of these intrinsic difficulties, the relevant message is that validating the theoretical strength with a total absorption measurement is a more stringent test of the quality of a nuclear model in a particular region of the nuclide chart than only looking at half-lives. This can have important implications in calculations related to astrophysical processes.

Another example of astrophysical applications is related to the estimation of neutron capture (n, γ) cross-sections for very exotic neutron-rich nuclei. Neutron capture (n, γ) plays a key role in the r process of nucleosynthesis. The cross sections are very difficult to determine experimentally. Indeed in many cases they cannot be determined because of the challenge imposed by the targets required. Direct measurements will require very innovative approaches hence current efforts concentrate on indirect methods [91]. In these circumstances, the use of theoretical estimates remains the only valid option. Theoretical estimates of the cross sections are based on the statistical Hauser–Feshbach model [92] that uses average quantities, namely nuclear level densities, photon strength functions and neutron transmission coefficients. These are parameterised using data mostly related to nuclei close to stability and consequently there is considerable uncertainty on the values needed in regions that are important for the r process calculations. In this context, the β-Oslo method was proposed in [93] to obtain information on nuclear level densities and photon strength functions from TAS measurements in β-decay studies. The method relies on the assumption that the total deposited energy in the spectrometer is equivalent

---

[7] Prolate or oblate deformed nuclei are typically axially symmetric. Triaxial nuclei have unequal symmetry axes, which makes their description more complex.

to the excitation energy of levels populated in the decay, which is violated due to detector inefficiencies thus introducing an unknown bias in the results.

In [94, 95] a method to constrain experimental quantities that are required in the Hauser–Feshbach estimate for very exotic neutron rich nuclei was proposed. It is based on the analogy between radiative neutron capture reactions and the process of β-delayed neutron emission. On the one hand radiative neutron capture depends mostly on Γ_γ and weakly on Γ_n. On the other hand β-delayed neutron emission can provide the ratio Γ_γ/Γ_n assuming that we are able to measure the γ emission from neutron unbound states that is expected to be weak. This is where the sensitivity of the TAGS technique can play a role. The advantage of the method is that the measurements can be extended into regions quite far from stability.

The γ-neutron competition was studied for the ^{87,88}Br and ^{94}Rb decays [94, 95] and more recently for ^{95}Rb and ^{137}I [96]. In these measurements the P_γ of the decays, which characterises the γ-emission probability above the neutron separation energy S_n was determined. It should be noted that P_γ can be defined by analogy with the P_n value.[8] The results from these measurements show that in most of the cases P_γ is large, even larger than the P_n in some cases. The large P_γ for ^{137}I obtained in our measurements is consistent with the TAGS measurement of [51]. The reason for this surprising P_γ result is because of the nuclear structure of the nuclei involved in the decay chain. A large mismatch between the spins and parities of unbound states populated in the daughter nucleus and those of the available states in the final nucleus after n emission means that neutron emission is hindered by the centrifugal barrier. Other recent measurements have also found large P_γ values in the decay of ^{70}Co [97] and ^{83}Ga [98] and different nuclear structure effects were invoked to explain it. This notable result warns us about the neglect of γ-neutron competition in theoretical estimates of P_n mentioned earlier in the chapter (see also [99]).

Summarising, the TAGS technique can be relevant for many applications that require a proper determination of the feeding pattern of the beta decays involved, in particular for those applications where detecting feeding at high excitation energy in the daughter nucleus is essential.

References

1. Duke, C.L., et al.: Nucl. Phys. A **151**, 609 (1970)
2. Hardy, J.C., et al.: Phys. Lett. B **71**, 307 (1977)
3. Raman: Atomic Data and Nuclear Data Tables **78**, 1 (2001)
4. Neugart, R., Neyens, G.: Nuclear moments. In: Lecture Notes in Physics, vol. 700, p. 135 (2006)

---

[8] P_γ is defined as the gamma intensity emitted per 100 beta decays above the neutron separation energy.

5. Hammamoto I., et al., Hamamoto, I., Zhang, X.Z.: Z. Phys. A **353**, 145 (1995)
6. Sarriguren, P., et al.: Nucl. Phys. A **635**, 55 (1998); Sarriguren, P., et al.: Nucl. Phys. A **658**, 13 (1999); Sarriguren, P., et al.: Nucl. Phys. A **691**, 631 (2001)
7. Petrovici, A., Schmid, A., Faessler, A.: Nucl. Phys. A **665**, 333 (2000)
8. Nilsson, S.G.: Binding states of individual nucleons in strongly deformed nuclei. Doctoral Thesis (1955)
9. Moreno, O., et al.: Phys. Rev. C **106**, 034317 (2022)
10. Nacher, E., et al.: Phys. Rev. Lett. **92**, 232501 (2004)
11. Poirier, E., et al.: Phys. Rev. C **69**, 034307 (2004)
12. Heyde, K., Wood, J.L.: Rev. Mod. Phys. **83**, 1467 (2011); Wood, J.L., et al.: Phys. Rep. **215**, 101 (1992)
13. Perez-Cerdan, A.B., et al.: Phys. Rev. C **88**, 014324 (2013)
14. Briz, J.A., et al.: Phys. Rev. C **92**, 054326 (2015)
15. Estevez Aguado, M.E., et al.: Phys. Rev. C **92**, 044321 (2015)
16. Rubio, B., et al.: J. Phys. G: Nucl. Part. Phys. **44**, 084004 (2017)
17. Algora, A., et al.: Phys. Lett B **819**, 136438 (2021)
18. Dombos, A.C., et al.: Phys. Rev. C **103**, 025810 (2021); Gombas, J., et al.: Phys. Rev. C **103**, 035803 (2021)
19. Ichimura, M., Sakai, H., Wakasal, T.: Prog. Part. Nucl. Phys **56**, 451 (2006)
20. Martinez-Pinedo, G., Poves, A., Caurier, E., Zuker, A.P.: Phys. Rev. C **53**, R2602 (1996)
21. Gysbers, P., et al.: Nat. Phys. **15**, 428 (2019)
22. Karny, M., et al.: Eur. Phys. J. A **25**(s01), 135 (2005)
23. Towner, I.S.: Nucl. Phys. A **444**, 402 (1985)
24. Burkard, K.H., et al.: Nucl. Instrum. Methods **139**, 275 (1978); Bruske, C., et al.: Nucl. Instr. Meth. **186**, 61 (1981); Burkard, K., et al.: Nucl. Instr. Meth. **126**, 12 (1997)
25. Karny, M., et al.: Nucl. Inst. Methods Phys. Res. B **126**, 411 (1997)
26. Plettner, C., et al.: Phys. Rev. C **66**, 044319 (2002)
27. Hu, Z., et al.: Phys. Rev. C **60**, 024315 (1999)
28. Algora, A., et al.: Phys. Rev. C **70**, 064301 (2004)
29. Algora, A., et al.: Phys. Rev. C **68**, 034301 (2003)
30. Cano-Ott, D.: Ph.D. thesis, University of Valencia (2000)
31. Nacher, E., et al.: Phys. Rev. C **93**, 014308 (2016)
32. Batist, L., et al.: Eur. Phys. J. A **46**, 45 (2010)
33. Algora, A., Rubio, B.: Studies of the beta decay of ^{100}Sn and its neighbours with a Total Absorption Spectrometer (TAS). RIKEN proposal NP1612-RIBF147 (2016); Algora, A., et al., RIKEN Progress Report (2019). p. 30
34. Hinke, C.B., et al.: Nature **496**, 341 (2012)
35. Lubos, D., et al.: Phys. Rev. Lett. **122**, 222502 (2019)
36. James, M.F.: J. Nucl. Energy **23**, 517 (1969)
37. Way, K., Wigner, E.: Phys. Rev. **73**, 1318 (1948)
38. JEFF and EFF projects. http://www.oecdnea.org/dbdata/jeff/. http://www.oecd-nea.org/dbdata/jeff/; Plompen, A.J.M., et al.: Eur. Phys. J. A **56**, 181 (2020)
39. Brown, D., et al.: Nucl. Data Sheets **148**, 1 (2018). https://www-nds.iaea.org/exfor/endf.htm
40. Shibata, K., et al.: J. Nucl. Sci. Technol. **48**(1), 1–30 (2011). https://wwwndc.jaea.go.jp/jendl/j40/j40.html
41. https://en.cnnc.com.cn/2020-06/17/c_501119.htm
42. Yoshida, T., et al.: J. Nucl. Sci. Technol. **36**, 135 (1999)
43. Nichols, A.L.: IAEA Report No. INDC(NDS) 0499 (2006); Gupta, M., et al.: IAEA Report No. INDC(NDS) 0577 (2010)
44. Äystö, J.: Nucl. Phys. A **693**, 477 (2001); Moore, I.D., et al.: Nucl. Instrum. Methods B **317**, 208 (2013)
45. Kolhinen, V., et al.: Nuclear Instr. Methods Phys. Res. A **528**, 776 (2004); Eronen, T., et al.: Eur. Phys. J. A **48**, 46 (2012)
46. Jordan, D.: PhD thesis, University of Valencia (2010)

47. Algora, A., et al.: Phys. Rev. Lett. **105**, 202501 (2010)
48. Jordan, D., et al.: Phys. Rev. C **87**, 044318 (2013)
49. Algora, A., et al.: Eur. Phys. J. A **57**, 85 (2021)
50. Fijałkowska, A., et al.: Phys. Rev. Lett. **119**, 052503 (2017); Rasco, B.C., et al.: Phys. Rev. C **95**, 054328 (2017)
51. Rasco, B.C., et al.: Phys. Rev. Lett. **117**, 092501 (2016)
52. Krzysztof Rykaczewski private communication
53. Kellet, M., Bersillon, O.: EPJ Web Conf. **146**, 02009 (2017)
54. Giot, L.: Private communication
55. Vogel, P., et al.: Phys. Rev. C **24**, 1543 (1981)
56. Hayes, A.C., Vogel, P.: Annu. Rev. Nucl. Part. Sci **66**, 219 (2016)
57. Double Chooz Collaboration: Nat. Phys. **16**, 556 (2020). https://doi.org/10.1038/s41567-020-0831-y
58. http://juno.ihep.cas.cn/Junojuno/Purjuno/201403/t20140306_117331.html
59. Ciuffoli, E., Evslin, J., Zhang, X.: Phys. Rev. D **88**, 033017 (2013)
60. Li, Y., Cao, J., Wang, Y., Zhan, L.: Phys. Rev. D **88**, 013008 (2013)
61. Schreckenbach, K., et al.: Phys. Lett. B **99**, 251 (1981); Schreckenbach, K., et al.: Phys. Lett. B **160**, 325 (1985); von Feilitzsch, F., Hahn, A.A., Schreckenbach, K.: Phys. Lett. B **118**, 162 (1982); Hahn, A.A., et al.: Phys. Lett. B **218**, 365 (1989)
62. Mampe, W., et al.: Nucl. Instrum. Methods **154**, 127 (1978)
63. Mueller, T.A., et al.: Phys. Rev. C **83**, 054615 (2011)
64. Huber, P.: Phys. Rev. C **84**, 024617 (2011)
65. Mention, G., et al.: Phys. Rev. D **83**, 073006 (2011)
66. Lasserre, T.: Private communication
67. Fallot, M., et al.: Phys. Rev. Lett. **109**, 202504 (2012)
68. Zakari-Issoufou, A.A., et al.: Phys. Rev. Lett. **115**, 102503 (2015)
69. Sonzogni, A.A., Johnson, T.D., McCutchan, E.A.: Phys. Rev. C **91**, 011301(R) (2015)
70. Dimitriou, P., Nichols, A.L.: IAEA report INDC(NDS)-0676, IAEA. Austria, Vienna (2015)
71. Greenwood, R.C., et al.: Nucl. Instr. Methods Phys. Res. A **390**, 95 (1997)
72. Tengblad, O., et al.: Nucl. Phys. A **503**, 136 (1989)
73. Yoshida, T., Tachibana, T., Okumura, S., Chiba, S.: Phys. Rev. C **98**, 041303(R) (2018)
74. Estienne, M., et al.: Phys. Rev. Lett. **123**, 022502 (2019)
75. An et al., F.P.: PRL **118**, 251801 (2017); Adey, D., et al.: PRL **123**, 111801 (2019)
76. Kopeikin, V., et al.: Phys. Atomic Nuclei **84**, 1 (2021); Kopeikin, V., et al.: Phys. Rev. D **104**, L071301 (2021)
77. Hayen, L., Kostensalo, J., Severijns, N., Suhonen, J.: Phys. Rev. C **100**, 054323 (2019)
78. Hayes, A.C., Friar, J.L., Garvey, G.T., Jungman, G., Jonkmans, G.: Phys. Rev. Lett. **112**, 202501 (2014)
79. Sonzogni, A.A.: Phys. Rev. Lett. **116**, 132502 (2016)
80. Weigand, M., et al.: Phys. Rev. C **92**, 045810 (2015)
81. Palmisano-Kyle, A., et al.: Phys. Rev. C **105**, 065804 (2022)
82. Skowronski, J., et al.: J. Phys. G: Nucl. Part. Phys. **50**, 045201 (2023)
83. Guerrero, C., et al.: Nucl. Instr. Methods Phys. Res. A **608**, 424 (2009)
84. Simon, A., et al.: Nucl. Instr. Methods Phys. Res. A **703**, 16 (2013)
85. Watson, D., et al.: Nature **574**, 497 (2019)
86. Horowitz, C.J., et al.: J. Phys. Nucl. Part. Phys. **46**, 083001 (2019)
87. Yokoyama, R., et al.: Phys. Rev. C **100**, 031302(R) (2019)
88. Kawano, T., et al.: Phys. Rev. C **78**, 054601 (2008)
89. Algora, A., et al.: Eur. Phys. J. A **57**, 85 (2021)
90. Kratz, K.L., Möller, P.: Private communication
91. Larsen, A.C., et al.: Prog. Part. Nucl. Phys. **107**, 69 (2019)
92. Hauser, W., Feshbach, H.: Phys. Rev. **87**, 366 (1952)
93. Spyrou, A., et al.: Phys. Rev. Lett. **113**, 232502 (2014)
94. Taín, J.L., et al.: Phys. Rev. Lett. **115**, 062502 (2015)

95. Valencia, E., et al.: Phys. Rev. C **95**, 024320 (2017)
96. Guadilla, V., et al.: Phys. Rev. C **100**, 044305 (2019)
97. Spyrou, A., et al.: Phys. Rev. Lett. **117**, 142701 (2016)
98. Gottardo, A., et al.: Phys. Lett. B **772**, 359 (2017)
99. Mumpower, M.R., et al.: Phys. Rev. C **94**, 064317 (2016)

Chapter 5
Future Perspectives

Abstract In this chapter future perspectives of the use of the technique will be briefly presented.

In any field of research, unexpected breakthroughs can occur and, by their nature, they are always difficult to foresee. In consequence it cannot be expected that the ideas that follow can cover all possible outcomes.

If we take a conservative view, many new results in the application of total absorption spectroscopy will come as a natural development of the new radioactive beam facilities being upgraded or under development. These developments will push the limit of the accessible nuclei for study both on the neutron-deficient and the neutron-rich side of the nuclide chart. The continuously upgraded RIKEN RIBF facility in Japan [1], as well as facilities like FRIB in the USA that have just started to function at the time of writing, [2], SPIRAL2 in France [3], SPES in Italy [4] and FAIR in Germany [5] are examples of such facilities. The unprecedented beam intensities of the new facilities will allow access to a much wider range of exotic nuclei. Accordingly these developments will allow us to perform total absorption studies of very exotic species with enough statistics in the future. Obvious examples of possible future studies are the most exotic N = Z isotopes, that are hardly accessible today, and very neutron-rich nuclei. On the neutron rich side, it is very clear that beyond a certain point neutron emission will take over, but there will be a critical region before reaching this point, when the Q-values are large, that TAGS data free of *Pandemonium* will be very important. Moreover, in the region where gamma and neutron emission compete, TAGS data are essential to obtain the complete beta decay spectra, as we have seen in the previous chapter. Even in the case of dominating beta decay neutron (or several neutrons) emission, a reduced version of TAGS can be important to determine if the neutrons proceed to the ground state or to excited states in the daughter, by measuring the gamma decay. Another interesting aspect not yet fully exploited with the TAGS is the detection of gamma decaying isomers. They are very important in nuclear structure studies and in particular near closed shells. Apart from the nuclear structure impact of these studies, they will also improve the description of beta decays that are relevant for astrophysical processes by providing better data to test nuclear models as discussed earlier.

© The Author(s), under exclusive license to Springer Nature Switzerland AG 2024
A. Algora et al., *Total Absorption Technique for Nuclear Structure and Applications*,
SpringerBriefs in Physics, https://doi.org/10.1007/978-3-031-58864-8_5

Some of the above mentioned new facilities are in-flight separation facilities. In studies at such facilities we face challenges related to the purity of the beams in terms of charge states and isomers. However a technical solution is available involving the combination of in-flight separation and isotopic separation using gas cells and the multi-reflection time of flight technique (MRTOF). Such developments already exist at GSI and RIKEN [7, 8], and are planned for the SPIRAL2 facility in France. This will solve some of the inherent isotope separation difficulties of the in-flight fragmentation technique. On the other hand the continuous upgrading of the ISOL facilities like the ISOLDE facility at CERN [6] and the IGISOL facility in Jyväskylä will continue to provide radioactive beams of high quality and excellent purity over a wider range of elements.

Another possible future development is related to the use of new scintillating materials for constructing detectors. This could improve the overall energy resolution and at the same time provide very high γ detection efficiencies. For the moment this is technically possible, but economically very challenging. On the analysis side, continuous developments are also foreseen and are currently taking place. The segmentation of the spectrometers will be more common and not exceptional. As shown in this book, beta decay studies are relevant on their own, but on many occasions they also provide information that is complementary to other studies such as the determination of nuclear shape or neutrino physics. And in general in those applications where it is important to have decay data free from the *Pandemonium* effect, the total absorption technique will be the main determinant and thus essential.

References

1. https://www.nishina.riken.jp/ribf/
2. https://frib.msu.edu
3. https://www.ganil-spiral2.eu/scientists/ganil-spiral-2-facilities/accelerators/
4. https://www.lnl.infn.it/en/spes-2/
5. https://www.gsi.de/en/researchaccelerators/fair
6. https://isolde.cern
7. Ayet San Andrés, S., et al.: Phys. Rev. C **99**, 064313 (2019)
8. Rosenbusch, M., et al.: Nucl. Inst. Methods Phys. Res. A **1047**, 167824 (2023)

Appendix

A.1   Spin and Parity of Nuclear States (J^π)

Spin (s) is primarily defined as the intrinsic angular momentum carried by elementary and composite particles. In the same way as the concept is extended from elementary particles to composite particles it can be generalised to characterise the states of atomic nuclei. The generalisation is done following the rules of the algebra of angular momentum. The spin j can have two components: an intrinsic part s and the orbital angular momentum part l, that is the quantum mechanical counterpart of the angular momentum of orbital revolution. Similarly parity (π) is a quantum number associated with a point reflection of the wave function of the system around the origin of the coordinate system. Wave functions that are unchanged by the parity transformation are described as even functions (parity $+1$), while those that change sign under a parity transformation are odd functions (parity -1).

Nuclear states are characterised by the excitation energy and the spin and parity. As an example we can consider the simplest nuclear shell model to predict the possible states of the system and fill the nucleons in those states according to both the Pauli and the minimum energy principles. Following this procedure we can determine theoretically the spin and parity of the lowest states. The ground states of even-even systems, where both Z and N are even, are characterised by zero spin $J = 0$ and even parity $\pi = +1$. Following this approach, since all nucleons are paired, the ground states of odd nuclei with one unpaired nucleon have the spin of the unpaired nucleon and parity $(-1)^l$, where l is the orbital angular momentum of the unpaired nucleon orbital. Odd-odd systems are more complex and the total spin of the nucleus in the lowest state is obtained as the vector sum of the angular momenta of the orbitals occupied by the unpaired nucleons and can take values between $|j_1 - j_2|$ and $|j_1 + j_2|$, where j_1 and j_2 are the spins of the unpaired particles. The parity is given by $(-1)^{(l_1+l_2)}$, where l_1 and l_2 are the orbital angular momenta of the unpaired proton and neutron respectively. For more details see [1].

© The Author(s), under exclusive license to Springer Nature Switzerland AG 2024   75
A. Algora et al., *Total Absorption Technique for Nuclear Structure and Applications*,
SpringerBriefs in Physics, https://doi.org/10.1007/978-3-031-58864-8

A.2   Gamma Radiation

Excited states in nuclei below the lowest threshold for break up into lighter fragments decay dominantly by electromagnetic transitions. In the γ emission process a nucleus changes to a state of lower excitation energy with the emission of a photon. The decay can also occur by internal conversion, where the energy of the transition is given to an electron in the atomic shell, which is emitted as described in the next section. The energy difference between the nuclear states accounts for the energy of the photon and the small recoil energy of the nucleus.

The intrinsic spin of the photon is $s = 1$, so the allowed J values carried by the electromagnetic transitions are $J = 1, 2, 3, \ldots$ where $J = 0$ is forbidden. The angular momentum conservation rule implies:

$$J_i + J_f \geq J \geq |J_i - J_f| \tag{A.1}$$

where J_i and J_f are the spins of the initial and final nuclear states and J is the angular momentum carried by the photon. Transitions between $J_i = 0$ and $J_f = 0$ are forbidden, but they can occur by internal conversion.

Parity is also conserved in electromagnetic transitions. Here we need to differentiate between *electric* transitions, in which the nucleus is coupled to the electric field of the photon, and *magnetic* transitions where the coupling occurs to the magnetic field of the photon. In the case of electric transitions the photon has parity $(-1)^J$, and in the case of magnetic type transitions the parity carried by the photon is $-1(-1)^J$.

Measurements of the energies of the photons provide information required to deduce the energies of excited states in the nucleus. The multipole character of the transitions, whether electric, magnetic or mixed and the spin carried, is usually determined from measurements of the angular distribution of the intensities and polarisation. This information is necessary to assign spins and parities to the nuclear states. For more detail see [2–4]. Because of the characteristics of the electromagnetic field, only the first allowed multipolarity and the next one in order are relevant for a particular transition as seen in the Table A.1.

Table A.1 Character of electromagnetic transitions with $|\Delta J|$ change of spin and π_i/π_f change of parity

π_i/π_f	$|\Delta J|$			
	0	1	2	3
−1	E1, M2	E1, M2	M2, E3	E3, M4
+1	M1, E2	M1, E2	E2, M3	M3, E4

A.3 Internal Conversion and Internal Conversion Coefficients (ICC)

Internal conversion is a de-excitation process of nuclear states where the excitation energy of the transition is given to an electron of the atomic shell and no gamma transition occurs. In the quantum mechanical model of an atom, there is a non-zero probability of finding an electron within the nucleus, so there is a certain probability that the nuclear transition energy can be given to an electron, most probably but not necessarily to electrons of the K shell since their orbit lies closest to the nucleus. Internal conversion is a non-radiative process and during the process the atomic number A does not change, i.e., there is no element transmutation, since the electron is emitted from the atomic shell and not by the nucleus as in the β decay process. Internal conversion is possible when the atomic nucleus is not fully ionised. The vacancy generated by the process is filled by other electrons with the subsequent emission of X-rays and Auger electrons. The process can compete with gamma emission with the exception of cases where γ transitions are not possible, i.e., $0^+ \rightarrow 0^+$ transitions. In those cases internal conversion is the dominant de-excitation process.

Conversion electrons are mono-energetic electrons emitted with an energy of:

$$E_{IC} = \Delta E - E_B \qquad (A.2)$$

Where E_{IC} is the conversion electron energy, ΔE represent the energy of the transition $E_f - E_i$, equivalent to the E_γ when gamma transitions are possible, and E_B represents the binding energy of the atomic electron.

To characterise the competition between gamma and internal conversion the internal conversion coefficient (*ICC* or α) is defined as:

$$\alpha_{tot} = I_e / I_{gamma} \qquad (A.3)$$

Defined as the quotient of the rate of emitted electrons and emitted gammas in the de-excitation process. α_{tot} is the sum of the contribution of the different atomic shells:

$$\alpha_{tot} = \alpha_K + \alpha_L + \alpha_M + \cdots \qquad (A.4)$$

where the α_n represents the contribution of the different atomic shells K, L, M, etc.

The conversion electron coefficient depends on the transition multipolarity, defined by the change in angular momentum and parity of the states involved in the transition. It is observed that the *ICC* increases both with atomic number Z and decreasing transition energies ΔE. For more details see [2, 5].

A.4   Classification of Beta Decays

Beta decays can be classified experimentally using the ft values introduced in Chap. 1 (see [6]). They can also be classified according to selection rules.

In the discussion of the Fermi theory presented in Chap. 1 spin degrees of freedom were not considered in order to simplify the initial discussion, but they play an important role in the classification of the transitions. Selection rules for beta decay follow from the symmetry properties of the Hamiltonian of the transition and the wave functions that enter the matrix element calculation. It is beyond the scope of this book to discuss the details in depth and the reader is encouraged to look at text books, such as [7], that describe the theory in detail. Instead of a detailed description a qualitative picture can be provided.

Initial and final states with different parities will lead to small matrix elements, where the Hamiltonian conserves parity, because of the reduced spatial overlap. So, from this perspective, parity conservation is a necessary requirement for having large matrix elements.

In quantum mechanics angular momentum is quantified as follows:

$$|l| = \hbar\sqrt{l(l+1)} \qquad (A.5)$$

where $l = 0, 1, \ldots$.

In the classical picture $l = r_o p$, where p is the linear momentum and r_o is the impact parameter of the particle. If we equate the two expressions in a "semi-classical" picture, we can obtain:

$$r_o = \frac{\hbar}{p}\sqrt{l(l+1)} \qquad (A.6)$$

In the beta decay process the leptons are emitted in the nucleus, so $r_o < R$, where R is the nuclear radius.

We can estimate for example what is the expected impact parameter r_o for a beta decay transition releasing 1 MeV.

$$r_o = \frac{\hbar}{p}\sqrt{l(l+1)} = \frac{\hbar c}{E}\sqrt{l(l+1)}$$
$$= \frac{197.3 MeV fm}{1 MeV}\sqrt{l(l+1)} = 197.3\sqrt{l(l+1)} fm$$

So, l should be zero for the decay to occur inside the nucleus, since the nuclear radius for any nucleus is of the order of a few fm.

This qualitative description shows that the emission of the leptons (ν, e) with $l \neq 0$ is improbable, and it leads to forbidden decays. If we define that allowed transitions should carry $l = 0$, then we can have two options:

Table A.2 Classification of Kth-forbidden unique beta-decay transitions [7]. ΔJ represents the spin change and $\pi_i \pi_f$ is the multiplication of the initial and final parities of the states involved in the transition

K	1	2	3	4	5	6
ΔJ	2	3	4	5	6	7
$\pi_i \pi_f$	−1	+1	−1	+1	−1	+1

Table A.3 Classification of Kth-forbidden non-unique beta-decay transitions [7]

K	1	2	3	4	5	6
ΔJ	0,1	2	3	4	5	6
$\pi_i \pi_f$	−1	+1	−1	+1	−1	+1

- Fermi transitions: the leptons are emitted with antiparallel spins

$$s_e(\uparrow) + s_\nu(\downarrow) + l_e + l_\nu = 0 \tag{A.7}$$

where $s_e(\uparrow)$ and $s_\nu(\downarrow)$ are the electron and neutrino intrinsic spins and l is their orbital angular momentum. In these transitions the spin orientation of the transforming nucleon is kept, and the total angular momentum of the nucleus does not change $\Delta J = 0$.
- Gamow–Teller (GT) transitions: the leptons are emitted with parallel spins:

$$s_e(\uparrow) + s_\nu(\uparrow) + l_e + l_\nu = 1 \tag{A.8}$$

So, in this case a spin flip takes place. The total angular momentum can change by $\Delta J = 0, 1$ excluding $0 \rightarrow 0$ transitions.

Allowed transitions can be of Fermi or Gamow–Teller type or both types simultaneously. For example the beta decay of the free neutron is a mixed Fermi-GT transition.

Forbidden decays violate the selection rules. They are classified according to the following tables (see Tables A.2 and A.3) [7].

A.5 Inventory of Nuclides for Decay Heat Calculations

As mentioned in the section devoted to applications, to estimate the decay heat one needs to determine the inventory of nuclides N_i that contribute at time t. This is done by solving a linear system of coupled first order differential equations for all fission products N_i:

$$\frac{dN_i}{dt} = -(\lambda_i + \sigma_i \phi)N_i + \sum_j f_{j \rightarrow i} \lambda_j N_j + \sum_k \mu_{k \rightarrow i} \sigma_k \phi N_k + y_i F \tag{A.9}$$

where λ_i is the decay constant of isotope i, σ_i is the neutron capture cross-section for isotope i, ϕ is the neutron flux, $f_{j \to i}$ is the branching ratio for the decay of j into i, $\mu_{k \to i}$ is the production rate of i per one neutron capture in k, y_i is the fission yield of isotope i and F is the fission rate.

Solving this equation requires nuclear data including fission yields, capture cross-sections and decay branching ratios. The equation represents the complexity behind the estimation of the inventory of nuclei necessary to determine the decay heat in reactors.

A.6   Neutrino Oscillations

One of the applications of the TAGS technique is the study of beta decays that contribute to the primary neutrino spectrum from reactors through the summation method. This can be of relevance for experiments aimed at the study of neutrino oscillations.

In the standard model of particle physics it is assumed that there are three neutrino types denoted by $|\nu_1\rangle$, $|\nu_2\rangle$, $|\nu_3\rangle$, having well defined masses. These states are the eigenstates of the mass operator and are orthogonal.

In the weak interaction neutrinos are produced in flavours. The corresponding eigenstates (electron ν_e, muon ν_μ, tau ν_τ) do not coincide with the mass eigenstates ($|\nu_1\rangle$, $|\nu_2\rangle$, $|\nu_3\rangle$). This means that when an electron neutrino is produced, this state is a linear combination of the mass eigenstates. This property is known as neutrino mixing.

For the sake of simplicity we discuss the case of a two component model (for more detailed descriptions see [4, 8]).

We begin by assuming that:

$$|\nu_e\rangle = cos\phi|\nu_1\rangle + sin\phi|\nu_2\rangle \tag{A.10}$$

then

$$|\nu_x\rangle = -sin\phi|\nu_1\rangle + cos\phi|\nu_2\rangle \tag{A.11}$$

represents flavour x and by construction is orthogonal to $|\nu_e\rangle$. This last state can represent a muon neutrino or a tau neutrino. The ϕ parameter is the mixing angle.

If the electron neutrino was created at $t = 0$ with linear momentum p, it can be represented by a plane wave $e^{ip/\hbar}$. At time t, the state can be represented by

$$|\nu_e\rangle(t) = e^{-iE_1 t/\hbar}cos\phi|\nu_1\rangle + e^{-iE_2 t/\hbar}sin\phi|\nu_2\rangle \tag{A.12}$$

where $E_1 = \sqrt{p^2c^2 + m_1^2c^4}$ and $E_2 = \sqrt{p^2c^2 + m_2^2c^4}$.

We can solve Eqs. A.10 and A.11 in terms of $|v_e\rangle$ and $|v_x\rangle$ and using (A.12) we can obtain:

$$|v_e\rangle(t) = e^{-iE_1t/\hbar}[(cos^2\phi + e^{i(E_1-E_2)t/\hbar}sin^2\phi)|v_e\rangle$$
$$-sin\phi cos\phi(1 - e^{i(E_1-E_2)t/\hbar})|v_x\rangle]$$

Using the rules of quantum mechanics the probability of detecting the neutrino as an electron neutrino after time t is the following:

$$P_e(t) = |\langle v_e|v_e\rangle(t)|^2 = |cos^2\phi + e^{i(E_1-E_2)t/\hbar}sin^2\phi|^2$$
$$= \quad 1 - sin^2 2\phi sin^2[(E_2 - E_1)t/2\hbar]$$

and the probability of detecting it as an x neutrino is

$$P_e(t) = |\langle v_x|v_e\rangle(t)|^2 = sin^2\phi cos^2\phi|(1 - e^{i(E_1-E_2)t/\hbar})|^2 = sin^2 2\phi sin^2[(E_2 - E_1)t/2\hbar]$$

So there is a certain probability that the electron neutrino will not be detected as such if $m_1 \neq m_2$ and consequently $E_1 \neq E_2$.

The oscillation pattern can be seem more clearly if we express t in terms of the distance travelled s and assume that the neutrinos are relativistic ($v \sim c$). For that we need to express $(E_2 - E_1)$ in momentum p using a Taylor expansion.

$$E_2 - E_1 = \sqrt{p^2c^2 + m_2^2c^4} - \sqrt{p^2c^2 + m_1^2c^4} \approx \frac{(\Delta m^2)c^4}{2pc} \quad \text{(A.13)}$$

where $\Delta m^2 = m_2^2 - m_1^2$.

Then

$$P_e(s) = P_e(t = s/c) = 1 - sin^2 2\phi sin^2[(E_2 - E_1)s/2\hbar c]$$

Using (A.13) we obtain:

$$P_e(s) = P_e(t = s/c) = 1 - sin^2 2\phi sin^2(\pi s/L) \quad \text{(A.14)}$$

where L, the so called oscillation length is defined as follows [4]:

$$L = \frac{4\pi(pc)(\hbar c)}{(\Delta m^2)c^4} = 2.48 \left(\frac{pc}{1MeV}\right)\left(\frac{1eV^2}{(\Delta m^2)c^4}\right) m$$

Because of the periodic dependence of (A.14) in s the phenomenon is called neutrino oscillation. Determining the mixing angles (in the more general case of three mixing angles), is one of the major goals of neutrino physics.

References

1. See for example Heyde, K.L.G.: The Nuclear Shell Model. Springer; Brussaard, P.J., Glaude-mans, P.W.M.: Shell-model Applications in Nuclear Spectroscopy. Nord-Holland Publishing Company
2. Krane, K.S.: Introductory Nuclear Physics. Wiley (1988). ISBN 0-471-80553-X
3. Muhin, K.N.: Experimental Nuclear Physics, vol. 1. Mir Publishers, Moscow (1987)
4. Cottingham, W.N., Greenwood, D.A.: An Introduction to Nuclear Physics. Cambridge University Press (1986). ISBN 978-0-521-651493
5. https://bricc.anu.edu.au
6. https://www.nndc.bnl.gov/nds/docs/NDSPolicies.pdf
7. Suhonen, J.: From Nucleons to Nucleus Concepts of Microscopic Nuclear Theory. Springer (2007). ISBN-10 3-540-48859-6
8. Hernández, P.: https://arxiv.org/pdf/1010.4131.pdf

Printed in the United States
by Baker & Taylor Publisher Services